Desert Dust in the Global System

A.S. Goudie   N.J. Middleton

Desert Dust in the Global System

With 114 Figures and 41 Tables

🐎 Springer

Prof. Dr. Andrew S. Goudie
St Cross College
St Giles
Oxford, OX1 3LZ
UK

Dr. Nicholas J. Middleton
School of Geography
Oxford University Centre for the Environment
South Parks Road
Oxford, OX1 3QY
UK

Cover illustration: A Seawifs image of a Saharan dust storm (see Fig. 5.9)

Library of Congress Control Number: 2006925945

ISBN-10 3-540-32354-6 Springer Berlin Heidelberg New York
ISBN-13 978-3-540-32354-9 Springer Berlin Heidelberg New York

Springer is a part of Springer Science+Business Media

springer.com
© Andrew S. Goudie and Nicholas J. Middleton 2006
Printed in Germany

Editor: Dr. Dieter Czeschlik, Heidelberg, Germany
Desk editor: Dr. Andrea Schlitzberger, Heidelberg, Germany
Cover design: Design & Production GmbH, Heidelberg, Germany
Production and typesetting: SPI Publisher Services

Printed on acid-free paper   31/3100   5 4 3 2 1 0

Acknowledgements

We are pleased to have worked with various colleagues over the years, including Richard Washington of the University of Oxford and Martin Todd of University College London. Our work has been greatly helped by the splendid web sites that make so much material available and we acknowledge the great stimulus to dust studies that has been provided by workers such as Jo Prospero. Sara Dickson (St Cross College, Oxford) kindly helped with the production of the manuscript, while Ailsa Allen of the Oxford University Centre for the Environment produced some of the figures with her customary skill and patience. Ben Hickey provided data on the Tokar Delta of Sudan and the Hamun Lakes of Afghanistan.

We are grateful to the following for permission to reproduce figures: Elsevier (Figs. 2.1, 3.3, 4.7, 7.5, 7.13, 7.15a, b, d, and 9.3), the Cambridge University Press, Cambridge (Fig. 2.5), Blackwell, Oxford (Fig. 3.3), Kluwer, Dordrecht (Fig. 6.1), the American Meteorological Society (Fig. 6.2), John Wiley and Sons, Chichester (Figs. 6.6, 10.7), *Annual Reviews* (Fig. 9.3), *Nature* (Figs. 7.7, 9.4), *Science* (Fig. 7.11), the Soil and Water Conservation Society (Fig. 7.4), the American Geophysical Union and the *Journal of Geophysical Research* (Fig. 7.15c), Cyril Moulin and The Institute Pierre-Simon Laplace (Fig. 5.8) and *Annales Geophysicae* (Fig. 7.7). We have also included selected illustrative material from our own previously published papers in *Earth Science Reviews*, the *Transactions of the Institute of British Geographers*, the *Annals of the Association of American Geographers*, *Climatic Change*, *Acta Universitatis Carolinae* (Prague) and the *Bulletin de la Classe des Sciences* (Académie Royale de Belgique).

Acknowledgements

Contents

1 The Nature and Importance of Dust Storms

1.1 Introduction

This book is about dust storms, atmospheric events that are typically associated with deserts. The study of desert dust, its entrainment, transport and deposition is an area of growing importance in investigations of global environmental change because dust storms have great significance for the physical environment and the world's human inhabitants (Table 1.1). Most dust events are generated by the erosion of surface materials in the world's drylands. Dry, unprotected sediments in any environment can be blown into the atmosphere, but the main sources of soil-derived mineral dust are located in desert regions. However, the impacts of wind-blown desert dust are global in their extent, making their study an area of major concern in Earth System Science.

Among the reasons why dust storms are important is that dust loadings in the atmosphere are significant for climate (Park et al. 2005). They affect air temperatures through the absorption and scattering of solar radiation (Haywood et al. 2003). In addition, dust may affect climate through its influence on marine primary productivity (Jickells et al. 1998); and there is some evidence that it may cause ocean cooling (Schollaert and Merrill 1998). Changes in atmospheric temperatures and in concentrations of potential condensation nuclei may affect convectional activity and cloud formation, thereby modifying rainfall (Bryson and Barreis 1967; Maley 1982) and possibly intensifying drought conditions.

Dust loadings may also change substantially in response to climatic changes, such as the North Atlantic Oscillation (Ginoux et al. 2004; Chiapello et al. 2005) or the Pacific Decadal Oscillation (Leslie and Speer 2005), to drought phases (Middleton 1985a; Littmann 1991a; Moulin et al. 1997; McTainsh et al. 2005) and in response to land-cover alterations (Tegen and Fung 1995). In these situations, the monitoring of dust storms can be indicative of environmental change.

Dust deposition provides considerable quantities of nutrients to ocean surface waters and the sea bed (Talbot et al. 1986; Swap et al. 1996). Aeolian dust contains appreciable quantities of iron (Zhu et al. 1997), the addition of which to ocean waters may increase plankton productivity (Gruber and Sarmineto 1997; Sarthou et al. 2003). Dust aerosols derived from the Sahara influence the nutrient dynamics and biogeochemical cycling of both terrestrial and oceanic

Table 1.1. Some environmental consequences and hazards to human population caused by dust storms

Consequence	Example
Environmental	
Algal blooms	Lenes et al. (2001a, b)
Butterfly transport	Davey (2004)
Calcrete development	Coudé-Gaussen and Rognon (1988)
Case hardening of rock	Conca and Rossman (1982)
Climatic change	Maley (1982)
Clouds	Sassen et al. (2003)
Coral reef deterioration	Shinn et al. (2000)
Desert varnish formation	Dorn (1986), Thiagarajan and Lee (2004)
Easterly wave intensification	Jones et al. (2003)
Glacier mass budget alteration	Davitaya (1969)
Loess formation	Liu et al. (1981)
Mercury translocation	Cannon et al. (2003)
Ocean productivity	Sañudo-Wilhelmy (2003), Jickells et al. (2005)
Ocean sedimentation	Rea and Leinen (1988)
Plant nutrient gain	Das (1988), Kaufman et al. (2005)
Playa (pan) formation and relief inversion	Khalaf et al. (1982)
Radiative forcing	Coakley and Cess (1985), Miller et al. (2004a)
Rainfall acidity/alkalinity	Stensland and Semorin (1982), Rogora et al. (2004)
Rock polish	Lancaster (1984)
Salt deposition and ground water salinization	Logan (1974)
Sediment input to streams	Goudie (1978)
Silcrete development	Summerfield (1983)
Soil erosion	Kalma et al. (1988)
Soil nutrient gain	Syers et al. (1969)
Stone pavement formation	McFadden et al. (1987)
Terra rossa formation	Delgado et al. (2003)
Tropospheric ozone	Bonasoni et al. (2004)
Ventifact sculpture	Whitney and Dietrich (1973)
Human-related	
Air pollution	Hagen and Woodruff (1973)
Animal madness	Saint-Amand et al. (1986)
Animal suffocation	Choun (1936)

Table 1.1. Some environmental consequences and hazards to human population caused by dust storms—cont'd

Consequence	Example
Asthma incidence	Gyan et al. (2005)
Car-ignition failure	Clements et al. (1963)
Closing of business	Gillette (1981)
DDT transport	Riseborough et al. (1968)
Disease transmission (human)	Leathers (1981)
Disease transmission (plants)	Clafin et al. (1973)
Drinking-water contamination	Clements et al. (1963)
Electrical-insulator failure	Kes (1983)
Machinery problems	Hilling (1969)
Microwave propagation	Ghobrial (2003)
Radio communication problems	Martin (1937)
Radio-active dust transport	Becker (1986)
Rainfall acid neutralization	Löye-Pilot et al. (1986)
Reduction of property values	Gillette (1981)
Reduction of solar power potential	Goossens and Van Kerschaever (1999)
Respiratory problems and eye infections	Kar and Takeuchi (2004), Chen et al. (2004)
Transport disruption	Houseman (1961), Brazel (1991)
Warfare	Agence France Press (1985)

ecosystems. Moreover, because of the thousands of kilometres over which the dust is transported, its influence extends as far a field as Northern Europe (Franzen et al. 1994), Amazonia (Swap et al. 1992) and the coral reefs of the Caribbean. Saharan dust has been suggested by Shinn et al. (2000) to be an efficient medium for transporting disease-spreading spores, which on occasion can cause epidemics that diminish coral reef vitality, a good match having been found between times of coral-reef die-off and peak dust deposition (Fig. 1.1). Atmospheric dust also influences sulphur dioxide levels in the atmosphere, either by physical adsorption or by heterogeneous reactions (Adams et al. 2005).

On land surfaces, additions of dust may affect soil formation. This has been proposed, inter alia, in the context of calcretes, salt horizons, *terrae rossae*, stone pavements and desert varnish (Thiagarajan and Lee 2004).

Dust additions play a major role in the delivery of sediments to the oceans (Fig. 1.2). For example, Guerzoni et al. (1999, p. 147) have suggested that: "Both the magnitude and the mineralogical composition of atmospheric dust inputs indicate that eolian deposition is an important (50%) or even dominant (>80%) contribution to sediments in the offshore waters of the entire Mediterranean

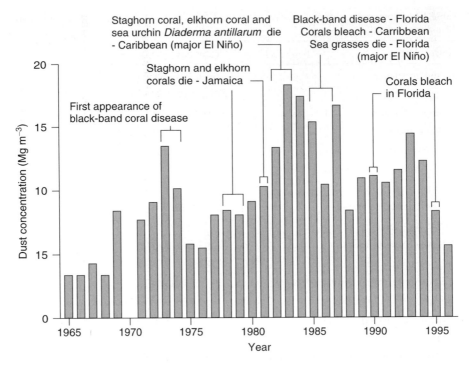

Fig. 1.1. The overall increase in dust reaching Barbados since 1965. Peak years for dust were 1983 and 1987. These were also the years of extensive damage to Caribbean coral reefs. Modified after Shinn et al. (2000)

basin". The role of dust sedimentation in the eastern Atlantic off the Sahara is also extremely important (Holz et al. 2004), and its significance in the Arctic Ocean has been discussed (Mullen et al. 1972; Darby et al. 1974).

Dust storms help to create various geomorphological phenomena by evacuating material from desert surfaces and then depositing it elsewhere. Desert depressions, wind-fluted bedforms (*yardangs*) and stone pavements are among such features. Above all, however, dust storms play a general role in the denudation of desert surfaces.

Dust storms also have many direct implications for humans. They can, for example, transport allergens and pathogens and disrupt communications. They may be a manifestation of desertification and of accelerated soil erosion. As 'Big Hugh' Bennett, father of the soil conservation movement in the United States, wrote at the end of the Dust Bowl: "To an alarming extent . . . the fertile parts of the soil are blowing away; to an equally alarming extent, menacing, drifting sand is left behind." (Bennett 1938b, p. 382)

Standard World Meteorological Organization (WMO) definitions for dust events that involve dust entrainment in the atmosphere are given by McTainsh and Pitblado (1987): (a) *Dust storms* are the result of turbulent winds raising large quantities of dust into the air and reducing visibility to less than 1000 m.

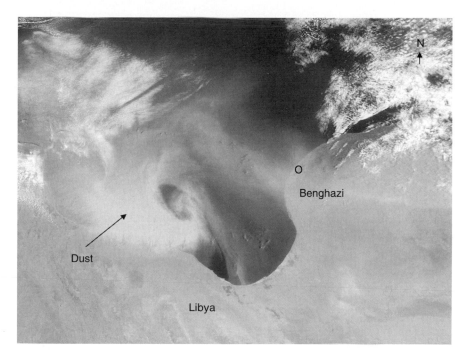

Fig. 1.2. Dust over northern Libya and the Gulf of Sirte, 26 May 2004 (MODIS)

(b) *Blowing dust* is raised by winds to moderate heights above the ground reducing visibility at eye level (1.8 m) but not to less than 1000 m. (c) *Dust haze* is produced by dust particles in suspended transport which have been raised from the ground by a dust storm prior to the time of observation. (d) *Dust whirls* (or *dust devils*) are whirling columns of dust moving with the wind and are usually less than 30 m high (but may extend to 300 m or more) and of narrow dimensions. There is some confusion in the literature between 'sand storms' and 'dust storms'. The former tend to be low altitude phenomena of limited areal extent, composed of predominantly sand-sized materials. Dust storms reach higher altitudes, travel longer distances and are mainly composed of silt and clay. In this work, the term *dust storm* refers to an atmospheric phenomenon in meteorology, where the horizontal visibility at eye level is reduced to less than 1000 m by atmospheric mineral dust.

While airborne particles in the world's atmosphere may be derived from a number of different sources – including cosmic dust, sea salt, volcanic dust and smoke particles from fire – in this book we concentrate very largely on the dust emitted from desert surfaces in low latitudes, though we recognize that dust may be emitted from glacial outwash material in polar regions and from disturbed agricultural land on susceptible soils in more humid parts of the world (Table 1.2).

Table 1.2. A selection of studies on wind erosion and dust deflation in non-desert regions

Region	Reference
Parts of Denmark	Møller (1986)
Swedish province of Skåne	Bärring et al. (2003)
Fenland and Breckland of eastern England	Goudie (1990, p. 302)
North-east of the Netherlands	Eppink (1982)
Northern Germany	Schäfer (1991)
Moravia and Silesia, Czech Republic	Hrádek and Švehlik (1995)
Southern Hungary	Mezösi and Szatmári (1998)
Southern Ukraine	Shikula (1981)
North-east Spain	López et al. (1998)
Parts of New Zealand	Marx and McGowan (2005)
Northern Canada	Nickling (1978)
Alaska	Péwé (1951)

1.2 Methods of Study

Desert dust has interested observers of the natural world for a very long time. Its transport over great distances has been noted in apparently bizarre depositional events such as 'blood rain' that are described in Homer's *Iliad* and in the works of numerous writers working in ancient Rome. Some of the earliest scientific observations were made by Charles Darwin (1846) off the west coast of Africa and Ehrenberg (1849) in the same area and in southern Europe, while von Richthofen's work in China was instrumental in establishing the aeolian origin of loess (von Richthofen 1882).

In contrast to this long history of reporting dramatic dust transport and deposition events, which has for the most part been largely descriptive (Fig. 1.3), it is only during the past few decades that aeolian dust has become a major environmental topic and that a more structured, systematic and quantitative approach to dust research has been developed (McTainsh 1999).

The study of dust storms has been carried out in a variety of ways. On the one hand, there are analyses that involve the long-term stratigraphic history of dust deposition in the oceans, in ice cores, in lakes and in loess sections. We return to this in Chapters 9 and 10. Archival studies have been undertaken, employing newspaper reports, diaries and the like. The classic study of this type is that undertaken for Kansas in the nineteenth century by Malin (1946). Then, there are studies that employ the analysis of observational data recorded at meteorological stations, using a set of standardized WMO Synop codes that relate to dust in the atmosphere (Table 1.3). This enables the frequency and distribution of dust storms to be mapped, though there are large

Fig. 1.3. A nineteenth century engraving of Saharan dust devils

tracts of the world's drylands where records are missing or imperfect. Current dust activity can also be monitored with ground- or air-based instruments such as lidar (Pisani et al. 2005), sun photometers, sun sky radiometers (Pinker et al. 2001; Masmoudi et al. 2003; Reid et al. 2003; Kaufman et al. 2005) and web cameras (Iino et al. 2004). The global Aerosol Robotic Network (AERONET) operated by the NASA Goddard Space Flight Center has been especially important in collecting near-real time data for a large number of sites globally (Kubilay et al. 2003). Dust can also be monitored and 'fingerprinted' Grousset and Biscaye (2005) to determine source areas by numerous means, including the analysis of mass size distributions, mineralogy, isotopic ratios, fossil content, plant waxes and pollen, and electron spin resonance (ESR; Table 1.4). Identification of source areas for specific long-range transport events can also be made using three-dimensional back-trajectory analysis for specific air masses (e.g. Betzer et al. 1988; Schwikowski et al. 1995; Kubilay et al. 2000).

Many devices have been developed to trap dust and measure the rate of its accumulation at the surface. Active samplers are equipped with pumping devices to maintain a flow through their intakes. They use filters of fine mesh (generally less than 2 μm) upon which particles accumulate. Small particle concentration can be monitored continuously at active sampling sites, using such devices as tapered element oscillating microbalances (TEOMs; see, for example, Kjelgaard et al. 2004; Xie et al. 2005). Passive samplers rely on wind

Table 1.3. WMO SYNOP present weather codes for dust events

Code figure ww	Symbol	Description		
05	∞	Haze		
06	S	Widespread dust in suspension in the air, not raised by wind at or near the station at the time of observation		
07	$	Dust or sand raised by wind at or near the station at the time of observation, but no well-developed dust whirl(s) and no duststorm or sandstorm seen		
08	⑧	Well-developed dust whirl(s) or sand whirl(s) seen at or near the station during the preceding hour or at the time of observation, but no duststorm or sandstorm		
09	(–S→)	Duststorm or sandstorm within sight at the time of observation or at the station during the preceding hour		
30	–S→|	Slight or moderate duststorm or sandstorm		_ has decreased during the preceding hour
31	–S→			_ no appreciable change during the preceding hour
32	|S→			_ has begun or has increased during the preceding hour
33	⩦S→|	Severe duststorm or sandstorm		_ has decreased during the preceding hour
34	⩦S→			_ no appreciable change during the preceding hour
35	|⩦S→			_ has begun or has increased during the preceding hour
98	⟁	Thunderstorm combined with duststorm or sandstorm at time of observation		

to maintain a flow through their intakes, but because they must use filters of much coarser mesh (generally greater than 40 μm), they are more suitable for sampling sand than dust. Moreover, passive samplers cause significant disturbance of the flow. This causes streamlines to diverge at the opening of the sampler; and dust particles tend to follow these streamlines rather than enter the collector. There are also various devices for measuring and sampling dry

Table 1.4. Methods used for dust monitoring and identification of source areas

Dust characteristic	Selected references
Mass size distribution	Prospero et al. (1970)
Mineralogy and elemental composition	Paquet et al. (1984)
Stable isotopes	Aléon et al. (2002), Wang et al. (2005b)
Lead isotopes	Turekian and Cochran (1981), Abouchami and Zabel (2003)
Rubidium–strontium isotopes	Biscaye et al. (1974)
Thorium isotopes	Hirose and Sugimura (1984)
Helium isotopes	Patterson and Farley (1997)
Neodymium isotopes	Grousset et al. (1998), Jung et al. (2004), Grousset and Biscaye (2005), Nakano et al. (2005)
Radon-222	Prospero and Carlson (1972)
Magnetic mineral assemblages	Oldfield et al. (1985)
Aluminium concentration	Duce et al. (1980)
Aerosol-crust enrichment	Rahn et al. (1981)
Rare earth element (REE) signature	Gaiero et al. (2004)
Single scattering albedo (SSA) signature	Collaud Coen et al. (2004)
Scanning electron microscopy of individual grain features	Prodi and Fea (1979)
Continentally derived lipids	Gagosian et al. (1981)
Pollen, plant waxes	Franzen et al. (1994), Dahl et al. (2005)
Enzyme activities	Acosta-Martínez and Zobeck (2004)
Trace elements	McGowan et al. (2005), Marx et al. (2005a, b)
Foraminifera	Ehrenberg (1849)
Electron spin resonance	Toyoda and Naruse (2002)

deposition fluxes, including bowls with or without water, buckets full of marbles or glass beads, moss bags, plastic mats with plastic straws like a grass lawn and inverted Frisbee samplers (Goodman et al. 1979; Hall et al. 1994; McTainsh 1999; Breuning-Madsen and Awadzi 2005). These tend to be cheap, simple and robust, but they are prone to contamination by bird excrement and the like; and different devices have differing capture efficiencies.

One tool that has become increasingly important in recent years for identifying, tracking and analysing large-scale dust events is remote sensing (Fig. 1.4). A range of different sensors has been used either singly or in combination (Table 1.5). These techniques give a global picture of dust storm activity, provide information on areas for which there are no meteorological station data, allow the tracking of individual dust plumes, enable sources of

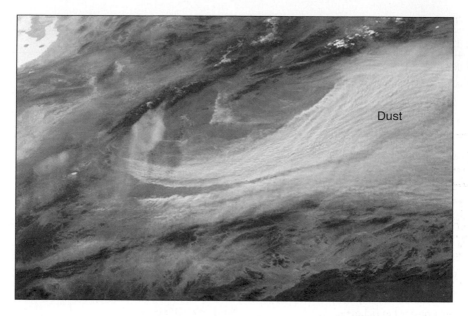

Fig. 1.4. A major dust storm in the Lut Desert east of Bam in Iran. The image was acquired by the crew of the International Space Station on 15 February 2004 (Earth Observatory, NASA)

dust to be precisely located and give information on such parameters as the optical thickness and altitude of dust.

Signals measured by satellite-based sensors generally include contributions from both the earth's surface and the intervening atmosphere, but a number of methods have been developed to identify that signal related to the radiative effect of atmospheric aerosols. These techniques include single- and multiple-channel reflectance, contrast reduction and polarization, multi-angle reflectance and thermal infrared emission (for a comprehensive review, see King et al. 1999). All of these approaches have their own drawbacks. Methods based on visible and infrared wavelengths, such as the Advanced Very High Resolution Radiometer (AVHRR) sensor carried on the National Oceanic and Atmospheric Administration's (NOAA's) polar-orbiting and Geostationary Operational Environmental Satellite (GOES), are adversely affected by clouds and water vapour and their use is restricted to either ocean or land surfaces. SeaWiFS (sea-viewing wide field of view sensor) is useful for detecting large plumes moving over the oceans but has difficulty detecting small and short-lived dust events over desert areas, due to their high radiance. Particular use has been made of the Total Ozone Mapping Spectrometer (TOMS) and the Moderate Resolution Imaging Spectroradiometer (MODIS). TOMS can detect UV-absorbing aerosols in the atmosphere, a method that does not suffer from the limitation of visible-wavelength techniques such as AVHRR because the UV surface reflectivity is low and almost constant over

Table 1.5. Examples of the use of remote sensing in the study of dust storms and dust aerosols

Sensor/satellite	References
LIDAR	Karyampudi et al. (1999), Chazette et al. (2001), Gobbi et al. (2002), Pisani et al. (2005)
METEOSAT	Legrand et al. (1994), Brooks (1999), Karyampudi et al. (1999), Brooks and Legrand (2000), Chazette et al. (2001), Chiapello and Moulin (2002), Leon and Legrand (2003)
MODIS: moderate resolution imaging spectroradiometer	Ichoku et al. (2004), Koren and Kaufman (2004), Jeong et al. (2005), Kaufman et al. (2005)
MISR: multi-angle imaging spectrometer	Zhang and Christopher (2003), Christopher et al. (2004)
TOMS: total ozone mapping spectrometer	Alpert et al. (2000), Alpert and Ganor (2001), Chiapello and Moulin (2002), Colarco et al. (2002), Ginoux and Torres (2003), Barkan et al. (2004), Mahowald and Dufresne (2004), Moulin and Chiapello (2004), Kubilay et al. (2005)
GOME: global ozone monitoring experiment	Guzzi et al. (2001), De Graaf et al. (2005)
AVHRR: advanced very high resolution radiometer	Husar et al. (1997), Cakmur et al. (2001)
AIRS: aqua advanced infrared radiation sounder	Pierangelo et al. (2004)
VISSR: visible and spin scan radiometer from fifth Japanese geostationary meteorological satellite (GMS-5)	Iino et al. (2004)
TMI: tropical rainfall measuring mission (TRMM) microwave imager	El-Askary et al. (2003)

both land and water. The TOMS UV spectral contrast data are, however, contaminated to a small degree by clouds and also suffer from an inability fully to detect aerosols within roughly 1–2 km above the surface (Mahowald and Dufresne 2004; Kubilay et al. 2005). Various recent studies have attempted to compare the results of different sensors with respect to measuring such parameters as aerosol optical thickness (AOT) or the Absorbing Aerosol Index (AAI; e.g. De Graaf et al. 2005; Jeong et al. 2005).

2 Dust Entrainment, Transport and Deposition

2.1 Introduction

Desert dust movement occurs in three phases: the entrainment or emission of material from the ground surface, its transport through the atmosphere and its deposition. These stages of wind erosion form the basis of this chapter, following an appraisal of the physical processes responsible for the formation of dust-sized particles and the geomorphological environments from which deflation typically occurs.

2.2 The Origin of Desert Dust Particles

Not all authorities agree on the upper grain-size limit for dust particles. Bagnold (1941) defines such particles as having diameters of less than 0.08 mm (80 μm), but many other workers prefer to define them according to the silt/sand boundary (i.e. less than 62.5 μm). Below this cut-off, fine particles are commonly categorised into those of silt and clay sizes, with grain diameters of 4.0–62.5 μm and <4.0 μm respectively (Wentworth 1922).

Whereas inorganic clay-size particles are generally agreed to be derived largely from chemical weathering, the processes responsible for silt formation in the desert environment remain a matter for debate. As Pye (1987) pointed out, many mechanisms of silt formation have been formulated but no clear picture regarding their relative importance has yet emerged. One major hypothesis is that silt can be formed by glacial grinding. This is an attractive theory to explain the great expanses of loess that occur on the margins of the former great Pleistocene ice caps (Smalley 1966; Smalley and Vita-Finzi 1968). Abrasion (sometimes called corrasion) during fluvial and aeolian transport may also produce silt. For example, numerous laboratory experiments have shown that abrasion of dune sand releases fines by spalling, chipping and breakage of particles and by the removal of grain surface coatings (Bullard et al. 2004; Bullard and White 2005). Moreover, many surfaces in both desert and polar regions show clear evidence of wind abrasion at a variety of spatial scales. In the latter case, some of the abrasion is achieved by driven snow, though snow abrasion is less efficient than that by quartz grain impacts. The greater kinetic energy of windblown sand compared to water transported sand explains the greater abrasion achieved by wind transport (Kuenen 1960).

Also of potential importance to silt formation are various types of weathering, including frost action, salt attack, thermal fatigue weathering and chemical weathering (see, for example, Goudie et al. 1979; Nahon and Trompette 1982; Smith et al. 2002). For instance, deeply weathered granitoid rocks may contain a quite high silt percentage – up to 37.7% in eastern Australia (Wright 2002). The role of salt weathering may also be important in producing what is often termed 'rock flour'. Goudie et al. (1979) designed an experiment to test whether silt-sized debris could be produced by salt weathering of aeolian dune sand, and found that it could. Subsequently other successful experimental simulations of salt attack on sands and on rocks were undertaken by Pye and Sperling (1983), Fahey (1985), Smith et al. (1987) and Goudie and Viles (1995). In addition, samples of salt-weathered rock collected in the field have shown that appreciable quantities of silt-sized material are produced (Goudie and Day 1980; Mottershead and Pye 1994).

Although the relative dominance of these mechanisms is difficult to assess, the important point to make is that silt can be produced in many ways, either singly, or more likely, in combination. Moreover, such mechanisms allow silt production in many types of environment, whether glacial, periglacial, arid or humid tropical (Wright 2001a; Smith et al. 2002). In addition, complex pathways of silt production and transport may be involved (Wright 2001b; Fig. 2.1). As Smith et al. (2002) remark:

"Weathering mechanisms coupled with periods of sediment reworking and associated silt production by glacial, fluvial and aeolian systems may provide a feasible explanation for the provenance of a significant majority of total global quartz silt. In addition to releasing silt-size particles directly, weathering may release considerable quantities of partially flawed sand grains. These flaws may then be readily exploited during subsequent periods of transport within glacial, fluvial or aeolian systems".

Some dust may be derived from erosion of organic materials (such as diatomite) which were deposited in pluvial lakes that have now become desiccated. Diatomite is a very light substance that, if abraded, produces fine, easily carried debris. This has been proposed as a major dust source in the Bodélé Depression in the Central Sahara (Giles 2005).

Other dust may be provided by the winnowing of fines from reactivated sand dunes. Dunes that have long been stable, having been produced under earlier conditions of greater aridity in, say, the Late Pleistocene, contain silt and clay contents in reasonably substantial quantities. Such fines may be the result of penecontemporaneous deposition of clay aggregates within the dunes as they were formed, but also important are post-depositional weathering and accretion of dust. Data from Kordofan (Sudan), north-west India, Zimbabwe, Niger and north-west Australia suggest that silt and clay contents of stabilized dunes can range from 7.8% to 32.0% (Goudie et al. 1993, Table 1).

a)

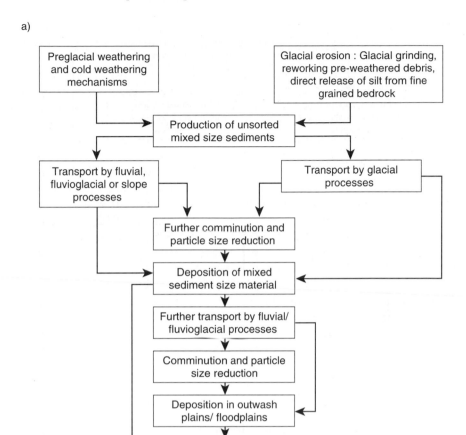

Fig. 2.1. a) Events in the formation of loess deposits – a hypothetical pathway to explain the formation of loess deposits associated with cold environments. Modified after Wright (2001b, Fig. 3)

b)

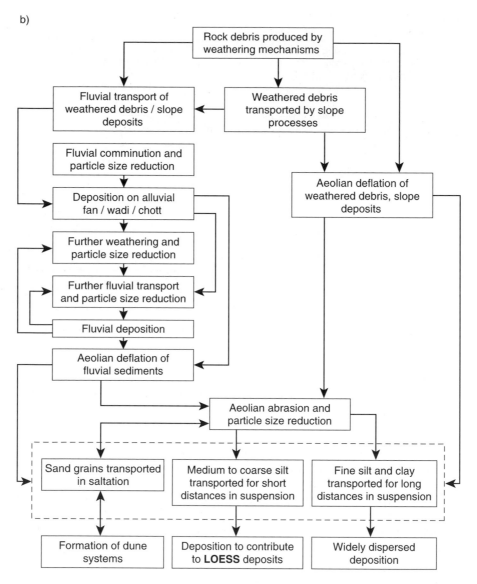

Fig. 2.1. (*Continued*) b) Events in the formation of loess deposits – a hypothetical pathway to explain the formation of loess deposits associated with hot environments. Modified after Wright (2001b, Fig. 4)

Thus, if such dunes become mobile as a result of climate or land-cover changes, they can release silt and clay for dust storm generation.

Given that there are so many mechanisms to produce silt-sized material (Smalley et al. 2005), it is not surprising that various geomorphological environments, in addition to old dunes, contain silt-sized material that is available

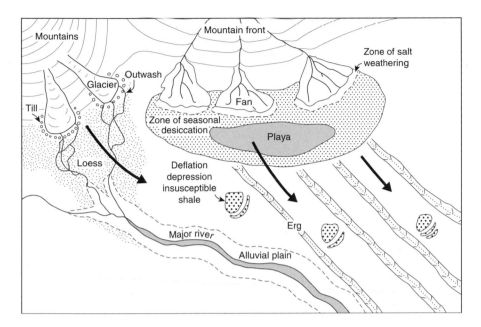

Fig. 2.2. A model of geomorphological environments from which substantial deflation occurs

for deflation. These include situations like outwash and alluvial fans, playa basins, weathered or unconsolidated rock exposures and areas of previously deposited loess (Fig. 2.2). Coudé-Gaussen (1984), whose work is largely based on the Sahara, has attempted to categorise desert surfaces that are highly favourable for producing dust:

- Dried-out salt lakes of internal drainage, the surface of which is disrupted and rendered mobile by salt crystallization
- Wadi sediments containing silt and the floodplains of great rivers, like the Niger
- Powdery areas (*fech-fech*) derived from ancient lake muds or on certain argillaceous rocks
- Desert clay soils (*takyrs*) with polygonal desiccation cracks
- Outcrops of rocks like unconsolidated Neogene fine-grained sediments

2.3   Threshold Velocities and Environments of Deflation

The threshold velocity is the minimum wind speed required to initiate deflation of surface sediments. At this velocity, the aerodynamic drag on the surface is enough to dislodge particles from the ground surface, to set them in motion and to lift them into the atmospheric boundary layer. The threshold velocity depends on a number of surface properties (see Section 2.4 below). The susceptibility of surfaces to deflation varies greatly, but very few sound empirical data are available to ascertain the critical threshold velocities for the input of soil particles into the air. A major attempt to rectify this problem was made by Gillette et al. (1980), employing a specially developed portable wind tunnel which permits the estimation of minimum threshold velocities under field conditions. They found that the velocities increase with different types of soil in the following sequence: disturbed soils, sand dunes, alluvial and aeolian sand deposits, disturbed playa soils, skirts of playas, playa centers, desert pavements. Table 2.1 (derived from Brazel 1991) indicates the threshold velocities for different types of surface in the American south-west. They range, according to material type, from 5.1 m s^{-1} to >16.0 m s^{-1}.

In addition, by studying the relationship between the occurrence of dust events and the wind speeds recorded by anemometers at meteorological stations, it is possible to see whether there is a characteristic wind speed at which dust is mobilized. In the Sahara, most dust-raising events are associated with winds between 6.5 m s^{-1} and 13.5 m s^{-1}, with a mean for all dust-raising events of 10.5 m s^{-1} (Helgren and Prospero 1987). Callot et al. (2000) found threshold values for the Sahara that ranged over 6.5–20.0 m s^{-1}, while for the Bodélé Depression in the central Sahara, Koren and Kaufman (2004) suggest a minimum threshold velocity of 10–11 m s^{-1}, with most of the values under 14 m s^{-1}. Lee et al. (1993) give an overall threshold value for the south High Plains of the United States of 6 m s^{-1}, while in China the threshold wind speed

Table 2.1. Wind threshold values for type surfaces in the United States South-West (after Clements et al. 1963; Nickling and Gillies 1989). From Brazel (1991)

Surface type	Threshold speed (m s^{-1})
Mine tailings	5.1
River channel	6.7
Abandoned land	7.8
Desert pavement, partly formed	8.0
Disturbed desert	8.1
Alluvial fan, loose	9.0
Dry wash	10.0
Desert flat, partly vegetated	11.0
Scrub desert	11.3
Playa (dry lake), undisturbed	15.0
Agriculture	15.6
Alluvial fan, crusted	16.0
Desert pavement, mature	>16.0

to generate a dust storm is generally considered to be between 6.5 m s^{-1} and 8.0 m s^{-1} (Kurosaki and Mikami 2005; Yabuki et al. 2005), though the values vary between different areas, with the Taklimakan Desert having values of 6–8 m s^{-1} and the Gobi Desert having values of 11–20 m s^{-1} (Laurent et al. 2005).

2.4   Wind Erosion of Soil and Other Surface Materials

Wind erosion occurs when the shear stress exerted on the surface by the wind exceeds the ability of the surface material to resist detachment and transport. Important controls of the susceptibility of soils to erosion include inherent properties of the soils themselves, including their grain-size characteristics, surface roughness and aggregate stability. The former includes clay content, which promotes cohesion, while the latter is greatly affected by soil organic content. It has long been recognized (Bagnold 1941; Chepil 1945) that the threshold velocity for particle movement increases as grain size increases, due to the effects of gravity, but that it also increases for the smallest particles, due to particle cohesion. The balance of these two effects produces an optimum particle size (ca. 60–80 µm) for which the threshold friction velocity is at a minimum. Land surface roughness is also a key factor. On the one hand, the threshold velocity required to initiate dust emission is increased in areas with higher surface roughness. On the other hand, the drag coefficient is also increased, leading to higher wind friction and thus to possibly higher dust emissions (Prigent et al. 2005).

Other important controls on a soil's erodibility include the degree of cover by non-erodible elements, such as rocks and vegetation (e.g. Merrill et al. 1999), and the moisture content, which affects the adhesive properties of the soil (Ravi et al. 2004). Snow cover (Kurosaki and Mikami 2004) will reduce wind erosion during winter months, though blowing snow can also break down soil aggregates. Seasonal freeze–thaw action is another way in which aggregate stability can be reduced (Bullock et al. 2001). Any surface crusts will also control rates of soil erosion (Singer and Shainberg 2004). Such crusts can be physical (e.g. clay skins, salt, lag gravels) or organic crusts composed of cyanobacteria, green algae, lichens and mosses. The importance of biological soil crusts for stabilizing arid zone soils and protecting them from wind erosion is becoming increasingly obvious (Belnap and Gillette 1998) and filamentous cyanobacteria mats are especially effective against wind attack (McKenna Neuman et al. 1996), partly because of their elasticity (Langston and McKenna Neuman 2005). However, these crusts are very susceptible to anthropogenic disturbance (Belnap and Gillette 1997). Table 2.2 illustrates the nature and direction of the effects on wind erosion of a range of soils, vegetation and landform conditions.

Numerous models have now been developed to predict wind erosion, with many of them developed from the prolific and influential pioneer work of

Table 2.2. Some key physical factors influencing wind erosion. *Symbols in parentheses:* + wind erosion becomes weaker; – erosion becomes greater as factor increases. Modified from Shi et al. (2004)

Climate	Soil	Vegetation	Landform
Wind speed (–)	Soil type	Type	Surface roughness
Wind direction	Particle composition	Coverage (+)	Slope (+)
Turbulence (–)	Soil structure		Ridge
Precipitation (+)	Organic matter (+)		
Evaporation (–)	Calcium carbonate (+)		
Air temperature (+)	Bulk density		
Air pressure (–)	Soil aggregation (+)		
Freeze–thaw action (+)	Soil water (+)		

Chepil and his co-workers (e.g. Chepil et al. 1962; Woodruff and Siddoway 1965). The Chepil wind erosion equation (WEQ) is:

$$E = f(I, C, K, L, V)$$

where E is the amount of wind erosion, I is a soil erodibility index, C is a local wind erosion climatic factor, K is a measure of local surface roughness, L is the maximum unsheltered distance across a field along the prevailing direction of wind erosion and V is the quantity of vegetation cover.

Subsequent models for predicting wind erosion include the Revised wind erosion equation (RWEQ) and the Wind erosion prediction system (WEPS; Visser et al. 2005).

Chepil and colleagues also devised a climatic index of wind erosion:

$$C = 100 \ U^3/(P - E)^2$$

where U is the average annual wind velocity at a standard height (10 m), and $P - E$ is the effective precipitation index developed by Thornthwaite (1948). This index assumes that wind erosion intensity varies with the cube of the wind velocity and the soil moisture content. McTainsh et al. (1990) also used a climatic index of potential wind erosion (E_w):

$$E_w = W(P - E)^{-2}$$

where W is the mean annual wind run (an indirect measure of wind velocity). They found that this simple index accounted for around two-thirds of the variance in dust storm activity in eastern Australia.

Some success has been gained by comparing dust emissions observed by satellite with predicted emissions based on analysis of wind velocities and the threshold conditions for dust emissions from mapped surface material types (e.g. Marticorena et al. 1999; Callot et al. 2000). Details of the Dust production model (DPM; developed by the LISA laboratory; University of Paris) which has two key parameters – aggregate size distribution and surface roughness – are provided by Lasserre et al. (2005) in the context of China.

Since Bagnold's classic work (Bagnold 1941), three modes of aeolian particle motion have been recognized: the rolling motion of the largest particles (*creep*), the hopping motion of particles in the size range ca. 50–500 µm (*saltation*) and the wafting of the smallest particles under the action of turbulent diffusion (*suspension*). The fraction undergoing suspension is dust, though saltation is a primary mechanism for the uplift of dust from the surface through a process called 'saltation bombardment' (Grini et al. 2002; Rampach and Lu 2004). Sand grains saltating over a surface of loose particles excavate ovoid-shaped micro-craters and a proportion of the material displaced from them is ejected into the flow. Saltation bombardment also breaks down aggregates.

There is some information to suggest that susceptible surfaces under appropriate climatic conditions can be deflated rather quickly. For example, the incision of wind-fluted bedforms (*yardangs*) into Saharan lake deposits that are of Neolithic pluvial age gives rates of deflation that are normally between 0.4 mm and 4.0 mm per year (Cooke et al. 1993). In the Kharga Oasis of Egypt (Fig. 2.3), yardangs almost 9 m high have developed in swamp deposits that were accumulating until ca. 4000 years ago, implying Late Holocene deflation of around 2000 mm ka^{-1} (Goudie et al. 1999). Boyé et al (1978) suggested that the Sebkha Mellala (Algeria) had been deflated at a rate of about 410 mm ka^{-1}, while Riser (1985), working in the Araouane Basin of

Fig. 2.3. A deflated yardang in the Western Desert of Egypt, which indicates the degree of deflation that has occurred in Holocene times (from ASG)

Mali, found a rate of 92 mm ka^{-1}. The Lop Nor yardangs in Central Asia may have been eroded since the fourth century AD, indicating a rate of wind erosion as high as 20 000 mm ka^{-1} (McCauley et al. 1977). Alluvium can also be deflated rapidly. In the Biskra region of Algeria, at least 1–4 m of deflation has occurred in less than 2000 years (Williams 1970, p. 61).

In general terms, it can be anticipated that soil surfaces disturbed by human activities may be especially susceptible to wind erosion and dust generation. Some studies have estimated that up to 50% of the current atmospheric dust load originates from anthropogenically disturbed surfaces (see, for example, Tegen and Fung 1995). However, a more recent study (Tegen et al. 2004) has suggested this may be an over-estimate and that dust from agricultural areas contributes <10% to the global dust load. Likewise, studies of dust over North Africa using the Infra-red difference dust index (IDDI) derived from METEOSAT (Brooks and Legrand 2000) suggest that there is little or no evidence that dust production is associated with widespread land degradation. Humans are responsible, however, in a variety of ways for generating 'fugitive dust', dust which escapes beyond the property line on which the source is located. Such dust comes from sources such as dirt roads, coal tips, mining sites, construction sites, stone crushers and sand- or gravel-processing plants.

2.5   Synoptic Meteorological Conditions Leading to Dust Events

Dust-raising events may occur under a wide variety of meteorological conditions within most global regions experiencing dust storms. However, the most frequent and severe dust storms typically occur under only one of a few synoptic meteorological conditions prevailing over any selected region. A number of dust-generating weather systems have been identified. By far the most important is the passage of low-pressure fronts with intense baroclinal gradients that are accompanied by very high velocity winds entraining and carrying dust. Surface cyclones themselves may sweep out gyres of dust, if circulation around the low pressure is sufficiently intense. In regions of monsoonal airflow, dust may be funneled along the convergence zone between cold air masses associated with regions of low pressure and hot, tropical anticyclonic air masses. More localized dust storms occur when katabatic winds (literally winds that blow downhill), such as the Berg winds of Namibia, deflate alluvial plains and fans adjacent to regions with considerable topographic relief. Convective plumes and vortices are active causes of dust-raising, and may contribute to about 35% of the global budget of mineral dust (Koch and Renno 2005).

At a local scale, dust devils and haboobs are significant for dust-raising and transport. Dust devils are highly localized rotating updrafts of buoyant air that develop over strongly heated surfaces (Fig. 2.4). Typical horizontal

Fig. 2.4. A large dust devil in Arizona (Courtesy of NASA)

velocities are about 10 m s^{-1}, their diameters are tens of metres and normally they persist for a matter of minutes (Warner 2004). They are visible because the horizontal wind speeds are sufficient to entrain surface dust and because the main upward motion in the outside of the vortex, combined with turbulence, causes the dust to rise. Quantitative field measurements have shown that the wind shears generated by dust devils are sufficient to lift all sizes of aeolian particles (Balme et al. 2003a). Haboobs, the name of which comes from the Arabic *habb*, meaning 'wind' or 'to blow', are convection-generated dust storms associated with thunderstorm activity. The colder outflow propagating ahead of a mature thunderstorm has high velocities (as large as 50 m s^{-1}) and a large vertical shear, which together may generate a dusty gust front

(Fig. 2.5). They are common, for example, in Sudan and Arizona. The dense wall of dust that is generated typically reaches a height of ca. 1000 m above the ground and the duration of the event tends to be a few hours.

We will now consider dust-raising conditions in the context of the main source regions, starting with the Southern Hemisphere. In Australia, dust storms generally follow the passage of strong low-pressure fronts tracking eastward across the south-eastern portion of the continent (Loewe 1943). The spectacular Melbourne dust storm of 1983 was generated by a non-precipitating cold front, ahead of which were extremely hot, low-level northerly winds. The frontal line represented a strong demarcation between a hot, north-west flow preceding the front and a west-south-west cooler flow following it (Shao 2000). Dust storms in February 2000, which transported dust from the Eyre Peninsula and New South Wales to New Zealand, were associated with a well developed summertime trough over the western half of Australia, preceded by a westerly trough-line associated with a surface level cold front and parent depression in the Southern Ocean. The trough-line marked the boundary between hot and dry pre-trough north-westerly airflow and colder westerly winds. The passage of such weather systems (McGowan et al. 2005) is associated with strong, turbulent surface winds, but with limited precipitation. During the summer monsoon, the convergence zone between high- and low-pressure systems may serve to channel dust from the interior of the Simpson Desert across Alice Springs and out over the Indian Ocean. Such a convergence may occur simultaneously with the movement of a low-pressure front

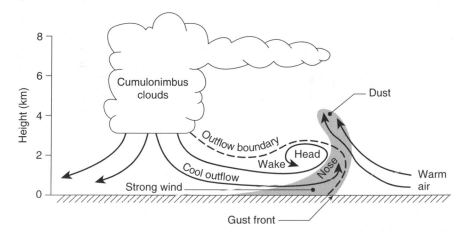

Fig. 2.5. Cross-section schematic of a haboob caused by the cool outflow from a thunderstorm, with the leading edge that is propagating ahead of the storm called an outflow boundary. The strong, gusty winds that prevail at the boundary are defined as a gust front. The leading edge of the cool air is called the nose and the upward protruding part of the feature is referred to as the head. Behind the roll in the windfield at the leading edge is a turbulent wake. The rapidly moving cool air and the gustiness at the gust front raise dust (*shaded*) high into the atmosphere. Modified after Warner (2004, Fig. 16.10)

across south-eastern Australia (Sprigg 1982). In addition, tropical cyclones which cross the northern coastline of Australia may generate dust as they track inland towards the dry interior (as with Hurricane Cecile in March 1984). In New Zealand, dust transport is associated with a range of conditions which include *föhn* winds, the passage of non-precipitating cold fronts and post-frontal south-westerlies (Marx and McGowan 2005).

In southern Africa, dust movement in Namibia tends to be caused by the *Berg winds*, a warm, dry, off-plateau, partially katabatic phenomenon. They occur primarily in winter, when a strong anticyclone occupies the interior and produces a strong outflow across the rim of the Great Escarpment down to the coastal plain. On the Andean Altiplano of Chile, north-western Argentina and southern Bolivia, dust is raised from *salars* and alluvial fans by superimposed westerlies across the region, with localized deflation accentuated by airflow around peaks. Dust is raised in the valleys of the Argentine foothills, especially in winter, by katabatic airflows known as *Zonda*. On the loessic plains of the Pampas (Wolcken 1951), the dust-raising winds are called the *Pampero*. They are caused by a low situated above Patagonia or the Falkland islands. Coming from the South Pacific, the cold front of this eastward-moving depression gives up much of its moisture on the western flanks of the Andes and then, when it meets the warm, humid air over the Argentinian plains, instability is created, which in turn creates a squall line of supercell thunderstorms.

When we turn to the Northern Hemisphere, in the mountainous regions of western North America, local katabatic winds generate dust storms in California (*Santa Ana* winds; Bowden et al. 1974) and along the Colorado Rocky Mountain Front. Summer haboobs during the 'Arizona Monsoon' are the primary dust-raising meteorological event in Arizona (Nickling and Brazel 1984; Brazel 1991) and may occasionally occur in the southern High Plains. Indeed, convective plumes and vortices lift large quantities of desert dust in the south-west United States (Koch and Renno 2005). In addition, low-pressure fronts tracking eastward may transport aeolian materials from agricultural regions of Texas and New Mexico into the Atlantic Ocean off the south-eastern coast (Henz and Woiceshyn 1980; McCauley et al. 1981). Surface cyclones crossing Texas may also raise dust palls. In addition, about 20% of the dust entrainment into the atmosphere over this region is associated with easterly wave activity.

In Morocco, Algeria and Libya, intense depressions may sweep bands of dust across the eastern Atlantic and central Mediterranean. Low-pressure fronts tracking across the North African coast carry dense dust palls to the Middle East (Yaalon and Ganor 1979). In Egypt, dust storms are associated with the passage of depressions and fronts tracking from western North Africa and across the Mediterranean (Banoub 1970). The convergence associated with the summer monsoon over eastern Africa channels the dust from Ethiopia, Somalia and northern Kenya across the Arabian Sea to the region north of Karachi. Farther north, in Sudan, a similar convergence creates dust storms crossing the Red Sea into Saudi Arabia, while the classic haboob of the Khartoum area remains the primary dust-moving system (Freeman 1952).

By far the most important global dust transport occurs with the passage of low-pressure fronts across the southern Sahara and Sahel. These depressions tend to track along a southerly course during the northern hemisphere winter, with more zonal easterly transport occurring during summer months. Throughout the year, trans-Atlantic export of dust from the Sahara may occur (Prospero and Nees 1977; Prospero et al. 1981), much of it within a well defined layer that extends up to altitudes of 5–7 km and is called the Saharan Air Layer (Prospero 1981). The role of easterly waves in dust entrainment and transport over north-western Africa is discussed by Jones et al. (2003, 2004). They argue that around 20% of the dust entrainment into the atmosphere over North Africa is associated with easterly wave activity.

The major meteorological conditions promoting dust storms in the Middle East are depressions moving eastwards from the Mediterranean across Turkey (Kubilay et al. 2005), the Levant (Michaelides et al. 1999) and northern Iraq. The *shamal* winds lifting dust from Iraq, Iran and adjacent regions (Fig. 2.6) are usually associated with low pressure anchored over southern Iran that forms a strong baroclinal gradient with a semi-permanent anticyclone over northern Saudi Arabia. The convergence zone between the two pressure systems induces high-velocity, turbulent winds for regional dust transport during a time of intense convection over the Tigris–Euphrates floodplain due to very high surface temperatures (Membery 1983).

Moving across Eurasia to the arid steppes and interior deserts of the former Soviet Union, low-pressure fronts following an easterly trajectory are again the primary agents of long-distance transport. Katabatic air flow may be locally important, such as the *Garmsil* wind that blows down the northern-facing slopes of Kopetdag, raising dust in Turkmenistan (Nalivkin 1983). Crossing the Hindu Kush and Karakoram Ranges, katabatic winds deflate the plains of the Indus and its tributaries and the Quaternary lakebeds and alluvial fans of Afghanistan and eastern Iran. These point-source dust storms can be characterized by extremely high-velocity surface winds and dense palls (Middleton 1986a). Within the Thar Desert of India and Pakistan, dust is transported by the westerly *Loo* wind in spring, the result of a strong pressure gradient brought about by a deepening of the seasonal trough, and haboobs, known locally as *andhi* (Joseph et al. 1980). Middleton and Chaudhary (1988) describe the dust storm of May 1986 in Karachi, which was generated by a thunderstorm associated with the passage of a monsoon depression.

Moving from Eurasia to China, low-pressure fronts transport dust aerosols over vast areas (Iino et al. 2004) and material may be carried in the upper westerlies to the Alaskan Arctic (Rahn et al. 1981) and into the Pacific Basin (Ing 1972). Local storms produced by katabatic winds may occur in the Tarim and Quaidam Basins, while upper-level westerly airflow probably generates point-source dust storms over the +4000 m Tibetan Plateau. Haboobs are known to occur in the Gobi Desert and are probably generated in the Kansu

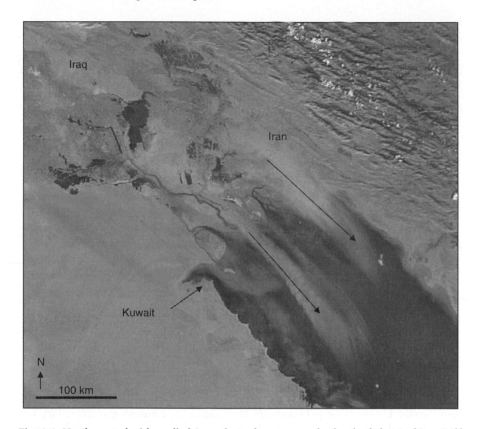

Fig. 2.6. North-westerly 'shamal' driven dust plumes over the head of the Arabian Gulf, 11 September 2004 (Seawifs)

region. The pervasive springtime dust events in China are largely driven by cold frontal systems (Aoki et al. 2005; Takemi and Seino 2005) connected with an upper-atmospheric trough located over Siberia and the north-eastern part of China, known as the East-Asian trough (Pye and Zhou 1989). This is associated with frequent and cold outbreaks from the north-west (Shao 2000). In Mongolia and northern China, the area of maximum dust storm generation is also associated with a zone of maximum negative vorticity, which induces a strong upward movement of air (Choi and Choi 2005).

2.6 Long-Range Transport

Most atmospheric dust falls back to earth a short time after entrainment and not far from its source, but dust storms are capable of transporting sediment over enormous distances, in many cases over some thousands of kilometres (Table 2.3). Dust from the Sahara is transported westwards to Amazonia, the

Table 2.3. Examples of long-distance dust transport

Approximate Distance (km)	Traced from	Traced to	Reference
6500	Sahara	Barbados	Delany et al. (1967)
8000	Sahara	Miami	Prospero (1981)
>1000	Sahara	Cape Verde Islands	Jaenicke and Schütz (1978)
2000	Sahara	Gulf of Guinea	Schütz (1980)
6500	Sahara	French Guiana	Prospero et al. (1981)
4000	Sahara	Berlin	MWR (1980)
7000	Sahara	Illinois	Gatz and Prospero (1996)
4000	Sahara	Hungary	Borbérly-Kiss et al. (2004)
7000	Sahara	Fennoscandia	Franzen et al. (1994)
10 000	Sahara	China	Tanaka et al. (2005)
750	Interior Morocco	Gibraltar	Ward (1950)
4000	Western Sahara	Cyprus	Gordon and Murray (1964)
2000	Libya and Egypt	Negev, Israel	Yaalon and Ganor (1975)
3500	Algeria	Denmark and USSR	VDL (1902)
700	Mkgadikdadi	Johannesburg	Resane et al. (2004)
10 000	Central Asia	Barrow, Alaska	Rahn et al. (1977), Andrews et al. (2003)
11 000	Central Asia	Tropical North Pacific (Eniwetok and Hawaii)	Turekian and Cochran (1981), Duce et al. (1980)
2000	West Kazakhstan	Baltic Sea	Hongisto and Sofiev (2004)
4000	China	Japan	Willis et al. (1980)
4000	China	Pacific Ocean (2500 km from coast)	Ing (1972)
>16 000	China	USA and Canada	Husar et al. (2001), McKendry et al. (2001)
>20 000	China	French Alps	Grousset et al. (2003)
>16 000	China	Greenland	Drab et al. (2002)
1500	Middle East	Southern USSR	Balakirev (1968)
3500	Caucasus	Rumania, Bulgaria and Czechoslovakia	Lisitzin (1972)
3500	Australia	New Zealand	Kidson and Gregory (1930)
3500	Australia	Singapore	Durst (1935)
2500	Canadian prairies	Illinois, USA	Van Heuklon (1977)
2500	Nebraska and Dakotas	Washington, D.C.	Hand (1934)
6000	Patagonia	Antarctica	Smith et al. (2003)
>7000	USA	Greenland	Smith et al. (2003)

Caribbean (Delany et al. 1967; Prospero et al. 1970), Bermuda (Chester et al. 1971) and the United States (Junge 1958). In Texas, Saharan events with moderate to high fine particulate contents occur on three to six days in the year, tend to be concentrated between June and August, last for one to three days and travel from their source in 10–14 days. Saharan dust also travels northwards to Europe, eastwards to the Middle East and even as far as China (Tanaka et al. 2005). Dust from Central Asia and China is regularly transported to Korea, Japan (Fig. 2.7), Hong Kong, the Pacific Islands and North America (Rahn et al. 1977; McKendry et al. 2001). Indeed, the frequency with which Asian dust reaches North America has probably been greatly underestimated and "contradicts the episodic characterization derived from short-term studies and anecdotal reports" (VanCuren and Cahill 2002). It has also been identified in snow pits at Summit in Greenland (Drab et al. 2002). The greatest distance desert dust particles have been found from their source is in excess of 20 000 km: dust from China has been identified as reaching the European Alps after being transported across the Pacific and Atlantic Oceans in some 315 h (Grousset et al. 2003).

Dust from the United States has been recovered from ice cores in Greenland and Patagonian dust from Antarctica (Smith et al. 2003). Material from Australian deserts crosses the Tasman Sea to New Zealand (Kidson and

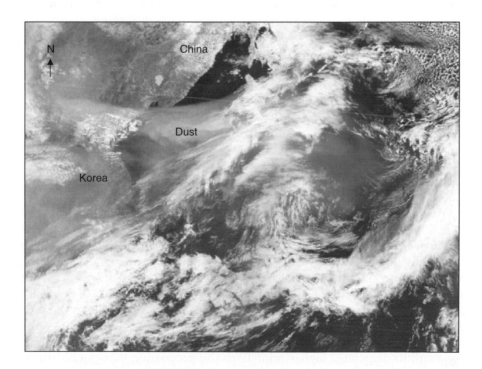

Fig. 2.7. Dust cloud over the Sea of Japan, 17 March 2002 (Seawifs)

Gregory 1930; Glaisby 1971; McGowan et al. 2000, 2005); and much dust from the Sonoran and Baja California deserts enters the eastern Pacific (Bonatti and Arrhenius 1965). Dust from the Caucasus settles in Romania, Bulgaria and Czechslovakia (Lisitzin 1972). We will treat the question of long-range transport in greater detail in Chapter 5.

2.7   Wet and Dry Deposition

The distance traveled by dust particles depends upon many factors, including wind speed and turbulence, dust grain characteristics and their settling velocities – the latter determined by the mass and shape of each particle. Atmospheric dust settles to the Earth's surface both through gravitational settling (dry deposition) and because of wet deposition with precipitation. Wet deposition can occur either below a cloud, when raindrops, snowflakes or hailstones scavenge dust as they fall, or within a cloud when dust particles are captured by water droplets and descend to earth when the precipitation falls. Wet deposition can sometimes be manifested in the phenomenon of 'blood rains'.

The relative importance of wet and dry deposition varies with the seasons, with rainfall amounts and with location. Wet deposition can be measured directly, but dry deposition is normally estimated by measuring aerosol dust concentrations and settling velocities (Prospero 1996b). A range of different methods is available, however, and these can give differing results (Torres-Padrón et al. 2002).

In the Mediterranean basin, dry deposition appears to be dominant, especially in the summer months, when typically the dust concentrations are at a maximum and rainfall amounts are low. The ratio of wet to dry deposition there is typically below 0.2 and the average about 0.1. By way of contrast, in the case of Asian dust deposition over the North Pacific, wet deposition exceeds dry deposition by up to a factor of ten (Zhao et al. 2003), whereas over interior China dry deposition dominates. Away from source, over Korea, Taiwan and the East China Sea, wet deposition dominates. Ginoux et al. (2004), using the Global ozone chemistry aerosol radiation and transport (GOCART) model, calculated that wet deposition accounted for 20.1% of total dust deposition over the North Atlantic, 10.0% over the South Atlantic, 33.3% over the North Pacific, 17.85% over the South Pacific, 22.56% over the North Indian and 20.0% over the South Indian. Over the Sahara at 21.25° N, wet deposition amounted to just 1.17%; and over the Sahel belt (<21.25° N) it amounted to 10.17%. The analyses by Torres-Padrón et al. (2002) of dust deposition in the Canary Islands, to the west of the Sahara, showed that the proportion made up by wet deposition varied between 3.4% and 8.6%. As one moves southwards to the belt more affected by the Intertropical Convergence Zone and its associated higher rainfall, so the proportion that can be attributed to wet deposition climbs (Sarthou et al. 2003).

Over land, dust is often subject to dry deposition when particles in suspension cross a boundary to terrain with a greater roughness. The presence of vegetation is thought to be important for trapping dust, while rock fragments also perform the same function, although such terrain probably retains less than 20% of settled dust (Goosens 1995).

2.8 The 'Giant' Dust Particle Conundrum

Generally the bigger a dust particle, the sooner it will fall back to earth after suspension; and the large majority of particles transported >100 km from source are <20 μm in diameter, in accordance with conventional theories on settling velocities (Gillette 1979). However, several workers have found sand-sized particles (>62.5 μm), or so-called 'giant' dust particles, in samples collected at considerable distances from source. 'Giant' Saharan dust particles have been noted in several locations: over the Cape Verde Islands (Glaccum and Prospero 1980), in Fuerteventura in the Canary Islands, in Corsica and southern France (Coudé-Gaussen 1989) and in southern Britain (Middleton et al. 2001). However, some similarly large particles have been recorded at even greater distance from source. Dust from north-east Asian deserts has been found >10 000 km out over the Pacific Ocean (Betzer et al. 1988).

Such large mineral grains are unexpected at such great distances from source because of their high fall velocities. Their aeolian mode of transport is undeniable (Betzer et al. 1988; Middleton et al. 2001) but these transport distances cannot be explained using currently acknowledged atmospheric transport mechanisms.

3 Environmental and Human Consequences

3.1 Introduction

As we saw briefly in the introduction to Chapter 1, the entrainment, transport and deposition of desert dust interacts with many other processes and forms in the physical world and has numerous implications for human societies. This is why the study of dust is becoming so significant in the burgeoning field of Earth System Science.

3.2 Marine Ecosystems

The movement of desert dust through the atmosphere is an important means by which numerous elements reach the oceans (Vink and Measures 2001; Fig. 3.1) and is thus of consequence for both their optical properties (Claustre et al. 2002) and their biogeochemistry, though large uncertainties about its effects remain (Jickells et al. 2005). Iron-rich dusts from the Gobi have been shown to cause a big increase in marine phytoplankton in the North Pacific (Bishop et al. 2002), while Saharan dust outbreaks provide an explanation for blooms of *Trichodesmium* (a filamentous diazotrophic cyanobacterium) on the West Florida Shelf (Lenes et al. 2001a, b). More generally Saharan dust, by supplying iron and phosphorus, promotes nitrogen fixation and hence oceanic primary productivity in the eastern tropical North Atlantic (Mills et al. 2004), the South Atlantic (Sañudo-Wilhelmy and Flegal 2003) and the Mediterranean. Guieu et al. (2002) suggest that Saharan dust outbreaks account for 30–40% of the total atmospheric flux of phosphorus in the western Mediterranean. Much of the aluminium flux to the Arabian Sea comes from dust deposition (Schüsssler et al. 2005) and Subba Row et al. (1999) have shown that dust from Arabia provides essential micronutrients for phytoplankton in the Arabian Gulf. Algal blooms (red tides) may also be triggered as a result of nutrient delivery by dust (for example, for the Arabian Sea, see Banzon et al. 2004). Dimethylsulfide (DMS) released from phytoplankton produces cloud condensation nuclei in the marine troposphere. This in turn increases cloud albedo and so can promote cooling of the atmosphere (Henriksson et al. 2000).

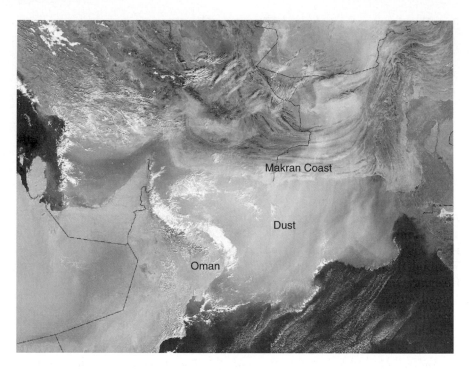

Fig. 3.1. A major dust event over the north-west Indian Ocean with the plume extending from Qatar over the Oman peninsula to the Rann of Kutch in north-west India, 13 December 2003 (SeaWifs)

Interestingly, however, not every outbreak of dust appears to generate a resulting increase in phytoplankton growth. Meskhidze et al. (2005) tracked two events that carried dust from the Gobi out over the Pacific and noted enhanced growth of phytoplankton after one event but not the other. They concluded that the difference was a function of the fact that the iron in desert dust is usually in a mineral form that has low solubility in seawater, hence it is not readily available to phytoplankton. These authors found that the dust event that did increase phytoplankton growth had been acidified by sulphur dioxide pollution from industrial plants in China, which had converted the iron to a more soluble form. Dust can also have an impact on ocean biogeochemistry by accelerating or inducing carbonate sedimentation by adsorption, ballasting and possibly aggregation of marine particles such as detritus or faecal pellets (Neuer et al. 2004).

Desert dust may contain living micro-organisms such as bacteria and fungi (Prospero 2004). The transport of dust from North Africa to the Caribbean has been implicated in the decline of corals in the region (Shinn et al. 2000; Garrison et al. 2003; Weir-Brush et al. 2004). The soil fungus, *Aspergillus sydowii*, which has been found in African dust samples, causes Black Band disease in a type of soft coral called the Sea Fan; and there appears to be a correlation between increased amounts of dust and the outbreak of the disease. Other diseases that may be related to dust are White Plague and White Pox. African dust containing pathogens could also be the cause of the widespread demise of reef-building staghorn corals and the sea urchin *Diadema*, which protects corals from being overgrown by algae. Dusts may also contain chemical contaminants which may alter the resistance of coral reef organisms to disease pathogens, affect reproduction or survival of larvae, interfere with calcification, or act as toxins (Garrison et al. 2003).

Dust blown from deserts and which settles on the sea floor may also have been involved in the formation of bedded sedimentary chert deposits (Cecil 2004). It may also have stimulated the growth of algal bioherms in the Late Paleozoic (Soreghan and Soreghan 2002).

3.3   Aeolian Erosion of Soils

As we saw in the last chapter (Section 2.4), dust storms result from the erosion and deflation of surface materials. This erosion has a number of consequences that can be classified into on-site and off-site effects (Goossens 2003; Table 3.1). The on-site effects include the preferential removal of fine particles. This leads to a gradual coarsening of topsoil, which is a cause of serious degradation for several reasons: soil nutrients are largely held by the fine particles and coarse sandy topsoil dries quickly.

More generally, extreme erosion can remove the entire surface soil, leaving behind sterile bedrock; and it can also remove soil organic carbon (Yan et al. 2005) and key nutrients (Masri et al. 2003). The eroded material may cause serious damage to crops and natural vegetation by abrasion (Woodruff 1956), a problem that can be particularly critical for young shoots when fields are poorly protected by vegetation cover. Young plants buried during dust storms can be adversely affected by the weight of the material deposited, consequent reduced photosynthesis and high soil temperatures during daytime. The resulting damage varies from a reduction in growth and development to a total destruction of crops, forcing farmers to resow their fields (Michels et al. 1993). Soil material lost from one area and subsequently deposited elsewhere may also contain potentially deleterious chemical residues, pathogens, weed seeds and the like. The off-site effects are dealt with more generally in this chapter.

3.4   Aeolian Contamination of Soils

The distinctive particle size and chemical constituents of dust, and the sometimes rapid rates at which dust accumulates, means that some soils owe much of their character to dust inputs. The contribution that dust makes to soil

Table 3.1. Some on-site and off-site effects of wind erosion (from Goossens 2003, Table 1)

On-site effects	Off-site effects
Soil degradation	**Short-term effects**
1. Fine material may be removed by sorting, leaving a coarse lag	1. Reduced visibility, affecting traffic safety
2. Evacuation of organic matter	2. Deposition of sediment on roads in ditches, hedges, etc.
3. Evacuation of soil nutrients	3. Deposition of dust in houses, on cars, washing, etc
4. Degrading water economy in the topsoil	4. Penetration of dust in machinery
5. Degrading soil structure	5. Deposition of dust on agricultural and industrial crops ruining their quality
6. Stimulated acidification of the topsoil	
	Long-term effects
Abrasion damage	1. Penetration of dust and its constituents in the lungs, causing lung diseases and other respiratory problems
1. Direct abrasion of crop tissue, resulting in lower yields and lower quality	2. Absorption of airborne particulates by plants and animals, leading to a general poisoning of the food chain
2. Infection of crops due to the penetration of pathogens	3. Deposition of heavy metals and other eroded chemical substances infecting the soil
3. Stimulated dust emission due to sand-blasting of the surface layer	4. Contamination of surface and groundwater via deposition of airborne particles
	5. Increased eutrophication of surface and groundwater
Other damage	6. Infection of remote uncontaminated areas, transforming these into new potential sources
1. Infection, with pathogens or soil constituents, of adjacent uncontaminated fields and crops	
2. Accumulation of low-quality wind-blown deposits on fields	
3. Building of sand accumulations at field borders, covering of drainage ditches	
4. Burial of plants	
5. Loss of seeds and seedlings	

profiles depends in part on topographic position. Goossens and Offer (2005) found that, in the Negev Desert of Israel, the highest rates of long-term accumulation occurred in valleys, especially those having a large catchment area, and on flat surfaces in a plateau position. Less, but still significant accumulation took place on concave windward slopes; and the lowest accumulation rates were on convex windward and lee slopes.

Information on dust characteristics is given in Chapter 6. Yaalon and Ganor (1973) introduced the term 'aeolian contamination' to describe the process by which soil properties have been modified by aeolian increments. They argued that the presence of significant amounts of quartz in soils derived from quartz-free substrates (e.g. basalts or limestones) could be indicative of such contamination. Since that time, numerous mineralogical studies have been undertaken which support this view: see, for example, Reheis (1990) on fan soils in Wyoming, Rex et al. (1969), Jackson et al. (1971) and Kurtz et al. (2001) on the lava soils of Hawaii, Naruse et al. (1986) on various soil types from Japan, Herwitz et al. (1996) on clay-rich palaeosols in Bermuda, Muhs et al. (1990) on the soils and bauxites of the Caribbbean, Vine (1987) on the ferralitic soils of southern Nigeria, Tiessen et al. (1991) on ferruginous soils in northen Ghana and Lee et al. (2004) on the soils of the South Shetland Islands (Antarctica). The *terra rossa* soils in southern Europe and the Levant (Yaalon and Ganor 1973; Mcleod 1980; Rapp 1984; Delgado et al. 2003) may also owe some of their features to aeolian accessions.

What is remarkable about such studies is their indication that soils at very substantial distances from desert margins are affected by dust, and not just those on the immediate desert margins (McTainsh 1984; Melis and Acworth 2001; Harper and Gilkes 2004). Thus, on *a priori* grounds, one might expect the soils in a dry continent such as Australia to show many types of soil in which aeolian deposition has played a role (Hesse and McTainsh 2003), including the clay-rich *parna*, but it comes as a surprise that recent studies have suggested that Saharan dust flux is crucial in Amazonia (Swap et al 1992; Kaufman et al. 2005) and that inputs of phosphorus derived from desert dust is vital for the maintenance of the long-term productivity of the rainforest (Okin et al. 2004).

In more general terms, desert dust can supply soils with many essential plant nutrients (e.g. Na, P, K, Mg), as well as substances that affect the availability of these nutrients (e.g. carbonates). This may stimulate the preferential growth of some plants over others, for example very saline dust may favour halophytes at the expense of other types (Blank et al. 1999). An assessment of the aeolian contribution to the fertility of soils on the Colorado Plateau (USA), where as much as 20–30% of surficial deposits comprise aeolian dust, found that the current plant community composition was heavily influenced by dust-derived nutrients (Reynolds et al. 2001). Dust inputs to the Colorado Plateau have enhanced the concentrations of P and Mo (both essential to nitrogen fixation) relative to bedrock values, P having doubled and Mo increased by a factor of 5. After identifying the minerals in atmospheric dust from ten widely scattered sites around the world, Syers et al. (1972) concluded that dust accessions can rejuvenate strongly leached and highly weathered soils. Feldspars, chlorites and micas brought in desert dust add K, Ca and Mg to soils over the long term.

Aggradation also plays a role in soil carbon sequestration, since the accumulation of dust buries the landscape and increases solum thickness. In the process, new soil organic carbon (SOC) is accumulated in the freshly deposited dust, while previously acquired SOC is buried below the shallow depth at which it originally formed. Some of this may persist for hundreds to thousands of years because of slow decomposition rates below the depth of greatest biological activity, especially under dry climatic conditions (Jacobs and Mason 2005).

Dust plays a fundamental role in the storage of water, particularly in rocky deserts, because its storage capacity is much larger than that of most desert lithosols.

3.5   Stone Pavements

Stone or desert pavements are a widespread surface type in arid regions and consist of an armour of coarse particles that overlies a profile containing a substantial content of fines (Fig. 3.2). Although the surface armour may be produced by a number of mechanisms (such as deflational or sheet flood removal of fines, or the vertical migration of coarse particles as a result of frost action, wetting and drying), recent studies have suggested that dust additions from above contribute substantially to their formation. Through processes such as rain-splash and surface wash, dust continually accumulates below coarse clasts, leading to the development of underlying vesicular horizons. The clasts, according to this model, have never been buried as was once assumed, but rise upward on a vertically accreting aeolian mantle (McFadden et al. 1987; Wells et al. 1987; Anderson et al. 2002). Gravel surfaces certainly appear to be effective at promoting dust accumulation (Li and Liu 2003; Li et al. 2005).

3.6   Duricrusts

The input of aeolian dust has been suggested as important to the composition and formation of several types of duricrust, a form of hardened surface crust or nodular layer found in many dryland situations. Calcretes, calcium carbonate-rich crusts that occur in arid and semi-arid areas, can form in many ways, but one of the key models is that they are produced by aeolian additions of dust which are translocated downwards and then accumulate in the soil

Fig. 3.2. A stone pavement in the Farafra oasis of Western Egypt. The vehicle has broken the dark armoured surface lag, exposing the finer grained, light-coloured material beneath. This material is then susceptible to deflation (from ASG)

profile (the *per descensum* model; Goudie 1983). Dust can contain significant amounts of calcium carbonate (Champollon 1965; Schlesinger 1985) and mass balance and strontium isotope studies have demonstrated its role (Chiquet et al. 2000) in Spain, in New Mexico (Capo and Chadwick 1999) and in other parts of the south-west United States (Mayer et al. 1988; Naiman et al. 2000).

Gypsum crusts (gypcretes) are another important component of surface materials in arid regions and, as with calcretes, *per descensum* models have received some support, although there are many possible mechanisms for their formation. It is probable that gypsum, deflated as dust from saline closed basins (pans, playas, etc.), accumulates down-wind and becomes consolidated into a pedogenic gypsum crust (Watson 1979), as demonstrated in Tunisia (Coque 1962), Australia (Chen et al. 1991) and the Namib Desert (Eckardt et al. 2001). The gypsum content of dust in southern Nevada and California ranges from 0.1% to 7.0%, equivalent to a flux of 0.02–1.5 g m^{-2} year^{-1} (Reheis and Kihl 1995).

Examination of the micromorphology of bauxite in Western Australia, together with mass balance equations, suggested to Brimhall et al. (1988) that the accumulation of dust derived from chemically mature soils could explain the development of such material. This finding challenged the prevalent view that bauxite was formed by simple *in situ* residual enrichment by weathering.

The study attributed most of the bauxite's Al and Fe, present in much higher proportions than could have been derived from the weathering of local bedrock, to additions of dust. Appropriately weathered surface materials were found to be exposed in various locations to the east of the Darling Range bauxite deposit investigated.

Brimhall et al. (1991) later applied the same approach to the study of a laterite in Mali, West Africa, and concluded that its composition, like that of the bauxite in Western Australia, had been determined by the nature of aeolian inputs. The study found that the weathering of local rocks had contributed only a minor fraction of the laterite's Al, Fe, Si and Au. The bulk of these elements was attributed to additions of strongly weathered material brought to the site as airborne dust.

3.7   Salinization and Acidity

In addition to contributing to the formation of calcrete and gypcrete, dust may lead to accumulation of more soluble salts in soil profiles and thus contribute to salinization (see Goudie and Viles 1997, p. 67). On the Red Sea coast of Sudan, aeolian dust consists of aggregates cemented by halite (sodium chloride) (Schroeder 1985); and large quantities of saline dust are being blown off the desiccating bed of the Aral Sea. The most comprehensive survey of dust additions of saline materials to desert surfaces is that undertaken in the western United States by Reheis et al. (1995). Reheis and Kihl (1995) monitored the salt content in dust in southern Nevada and California from 1984 to 1989 and found the average soluble salt content (excluding gypsum) ranged from 4% to 19%, equivalent to a salt flux of 0.3–2.4 g m^{-2} year^{-1}.

Dust that is rich in soluble salts and bases may be quite strongly alkaline. Calcitic dust has been shown to contribute not only to calcretes, as discussed above, but also to speleothems found in various cave sites (Goede et al. 1998; Frumkin and Stein 2004). In addition to reducing the incidence of acid precipitation, including snow (Roda et al. 1993; Avila et al. 1997; Avila and Roda 2002; Rogora et al. 2004; Delmas et al. 2005), such alkaline dust may also change the pH of soil layers through direct deposition and by reducing the acidity of precipitation. Dust collected from the Harmattan in Ghana, for instance, had pH values that were strongly alkaline, ranging over pH 8.0–9.4 (Breuning-Madsen and Awadzi 2005). Modaihsh (1997) found that dust from Riyadh, Saudi Arabia, averaged pH 8.9. Acid precipitation has long been regarded as a major environmental problem because of its adverse and diverse effects upon ecosystems. It is also implicated in building-stone decay. The acidity of precipitation may, however, be reduced by desert aerosols, which are often rich in calcium and other bases and are frequently alkaline. Recent studies in southern Europe have shown that the pH of rainfall has increased in some areas (Fig. 3.3) at the same time as Saharan dust incursions

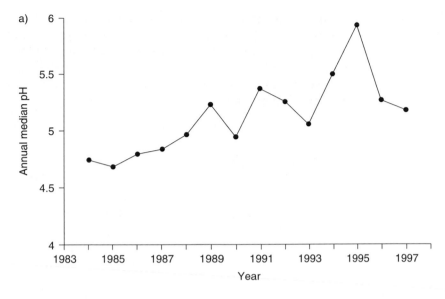

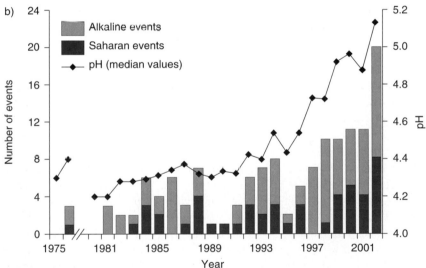

Fig. 3.3. Trends in pH of dust events over Europe. a) Evolution of the median pH of rain for 1983–1997 at Montseny, north-east Spain. The median pH is calculated for hydrologic years beginning on 1 August. Modified after Avila and Peñuelas (1999, Fig. 3). b) Number of alkaline and Saharan events at Pallanza, north-west Italy, since 1975 and the trend of median pH values. Modified after Rogora et al. (2004, Fig. 3)

have increased (see, for example, Avila and Peñuelas 1999; Rogora et al. 2004), though decreasing anthropogenic sulphate emissions over the same period may also have played a role. Nevertheless, significant inputs of Saharan dust have been suggested as a viable explanation for the fact that many low-alkalinity lakes in the Alps and the Pyrenees did not become acidic in the late twentieth century, unlike numerous lakes in areas rarely influenced by such dust depositions, for instance in Scandinavia (Psenner 1999). Given the important effects of desert dust on the chemical and nutrient balances in the oceans (see above), the study of similar impacts in freshwater bodies deserves much more attention than it currently attracts.

3.8   Desert Depressions and Yardangs

Arid regions are frequently characterized by large numbers of closed depressions (Fig. 3.4). This is particularly the case in the High Plains of the United States, the interior of Southern Africa, the Pampas and Patagonia in South America, the Manchurian and West Siberian plains and substantial parts of Australia (Goudie and Wells 1995). Although such depressions can result from a wide range of mechanisms (e.g. animal excavation, solution, tectonics), it has for long been proposed that many of them are caused by deflation (see, for example, Gilbert 1895) and that the production of fine-grained material by processes like salt weathering creates material that can then be removed in

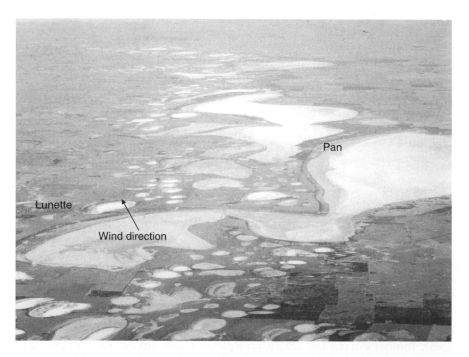

Fig. 3.4. An air photograph of a large series of pans (closed depressions) deflated out from old river channels in the interior of Western Australia (from ASG)

suspension downwind (see, for example, Du Toit 1906; Woodward 1897; Pelletier and Cook 2005). Closed deflation depressions also occur under cold climate conditions, where limited vegetation cover, surface disturbance by needle-ice formation and strong local winds can cause the excavation of suitable materials to occur (Seppälä 2004).

Yardang is a Turkmen word introduced by Hedin (1903) for wind-abraded ridges of cohesive material. Yardangs result from a number of formative processes, including wind abrasion, deflation, fluvial incision, desiccation cracking and mass movements (Laity 1994), but deflation is probably highly important in their formation and yardang areas are probably major sources of dust. They show a considerable range in scales, from micro-yardangs (small, centimetre-scale ridges), through meso-yardangs (forms that are some metres in height and length) to mega-yardangs (features that may be tens of metres high and some kilometres long; Cooke et al. 1993, pp. 296–297; Halimov and Fezer 1989; McCauley et al. 1977). These mega-yardangs are ridge and swale features of regional extent, called *crêtes* and *couloirs* in the French literature (Mainguet 1972).

The type site for yardangs is the Tarim Basin, for it is here that they were named by Hedin (McCauley et al. 1977). In his travels to Lop Nor, Hedin encountered these distinctive forms and called them yardang, the ablative form of the Turkestani word *yar*, which means ridge or steep bank. These yardangs appear to have developed in old lake and alluvial sediments. Major mega-yardangs also occur to the south-east of the Tarim Basin.

The Lut Desert of Iran contains classic mega-yardangs (Gabriel 1938) developed in Pleistocene basin fill deposits (silty clays, gypsiferous sands). The area involved is ca. 150 km long and 50 km wide. The ridges (*kaluts*) run from the north-west to south-east and attain heights of 60 m. They extend for tens of kilometres.

Mega-yardangs are extensively developed in northern Saudi Arabia, where they are formed in the Cambrian Sandstones and some other bedrocks. They are in excess of 40 m high and hundreds of metres long. Satellite images suggest that the bulk of them lie in an area extending over around 5° of latitude, which is bounded on the west by the marginal mountains or escarpment of the Red Sea Rift and on the east by the great Nafud Sand Sea. They appear to have been moulded by winds coming round from the west and west-south-west. The islands of Bahrain have small areas with large wind flutes. One area is developed on aeolianites (Jiddah Island), with yardangs 4–6 m high, while the other is developed on resistant Eocene limestones (Rus Formation) in the south-west corner of the main island's central depression. These latter features include aerodynamically shaped hills up to 10 m high, as well as larger hills that rise above the Central Plateau (Doornkamp et al. 1980, p. 200).

Northern Namibia is located in a hyper-arid area, with much of it underlain by ancient igneous and metamorphic rocks belonging to the Swakop Group (570–900 Ma). To the south of the Cunene sand sea, there is a very large area of wind-fluted basement rock that shows a great expanse of narrow, linear yardangs that trend approximately from south-south-east to north-north-west and appear to have similar orientations in that area to the barchans that move across their surface and to the orientations of the pre-dominant sand streams that have been identified in the Skeleton Coast sand sea to the south. The yardang area covers around 42 km by 25 km (ca. 1311 km²), with individual ridges running typically for distances of 8–10 km, with a spacing of around 300–350 m. In southern Namibia, between the Namib Sand Sea and the Orange River, there is a hyper-arid area with mega-yardangs developed in ancient crystalline and metamorphic rocks with complex structures. Many of the ridges are in excess of 20 km long and are ca. 1 km across. Some of the corrasional features near Pomona are 100 m high. There are at least four main areas where large yardangs occur: just to the south of Luderitz, near Pomona and inland from Chamais Bay.

The presence of vegetation-free surfaces, combined with the existence of strong, uni-directional winds from a northerly quarter, make it possible for wind-fluted surfaces to form in the Western Desert of Egypt. Yardangs are extensively developed, both in superficial materials and in bedrock. Yardangs were noted in the Western Desert by Bagnold (1933), who termed them 'mud lions'. Yardangs formed in playa sediments are widespread in the Dakhla depression (Brookes 1993) and in Farafra (Hassan et al. 2001), where the yardangs are up to 11 m high. Other yardangs occur on bedrock surfaces. Notable are those on the formations that cap the Libyan Plateau in the vicin-ity of Dakhla and Kharga (Brookes 1993). The yardangs develop best on those Tertiary limestones that do not contain a large content of chert. If chert is present, it armours the surface and lineated terrain is then replaced by smoooth chert-littered plains.

In the central Sahara, there are large areas of mega-yardangs, most notably in the Borkou region of Chad, to the north of Faya Largeau. Yardangs west of the Ounianga Kebir are commonly more than 20 km long, 1 km or more wide and separated by troughs ranging from 500 m to 2 km (McCauley et al. 1977, p. 50). Large yardangs occur in the far south of Algeria near the border with Mali and Niger (ca. 5° E, 20° N), in southern Algeria to the south of the Hoggar Massif (ca. 8° E 22° N) and also in an extensive area to the west of Tibesti. The features that occur to the south and west of Tibesti have been mapped by McCauley et al. (1977, Fig. 16). This is a major area of dust storm generation.

The High Andes of Latin America have extensive yardang fields. Those in Argentina have formed in ignimbrites or in lavas and show a general orienta-tion that is from north-west to south-east or from west-north-west to south-south-east. Most of the ridges are between 2 km and 10 km long.

Another classic area for mega-yardangs is the Peruvian Desert (McCauley et al. 1977). Although some occur in the Talara region of northern Peru, the most impressive forms occur in the Paracas-Ica Valley region of central Peru. They are intermediate in size between those of the Lut of Iran and those of the central Sahara. There is also an isolated area of yardangs on the coast of central Chile, near Chanaral (70° 43′ E, 26° 42′ S). They run from south-west to north-east and the largest are several kilometres long.

3.9   Dust and Radiative Forcing

Dust particles in the atmosphere exert both direct and indirect influences on climate. An example of the former is the effect that dust particles have on radiation budgets. Indirect influences include those brought about by the effects of dust on biogeochemical cycling (Moreno and Canals 2004) and, for instance, on carbon dioxide levels in the atmosphere. In addition, it needs to be remembered that the relationship between aeolian dust and climate is bidirectional, since climate plainly has a major impact on dust generation, transport and deposition. A specific illustration of this is that, in West Africa, easterly waves generate dust in the atmosphere, but the dust may also in turn lead to an intensification of easterly waves (Jones et al. 2004). Likewise it is also possible that radiative heating within a dust layer over Arabia reinforces the monsoon circulation which, through a positive feedback, raises additional dust into the atmosphere (Miller et al. 2004a).

Radiative forcing (the perturbation of the radiation balance caused by an externally imposed factor) by dust is complex (Tegen 2003), since it not only scatters but also partly absorbs incoming solar radiation; and it also absorbs and emits outgoing long-wave radiation (Li et al. 1996; Moulin et al. 1997; Alpert et al. 1998; Miller and Tegen 1998; Haywood et al. 2005). Changes in the amount of dust in the atmosphere would cause changes in the radiation balance and thus also in surface temperatures. However, the magnitude and even the sign of the dust forcing remains uncertain (Arimoto 2001), for it depends on the optical properties of the dust [which relates to its particle size, shape (Kalashnikova et al. 2005) and mineralogy], on its vertical distribution (Fouquart et al. 1987; Meloni et al. 2005), on the presence or otherwise of clouds (Quijano et al. 2000), on its moisture content (Kim et al. 2004) and on the albedo of the underlying surface (Nicholson 2000). Darker particles tend to absorb radiation and to scatter relatively little, so they may warm the air. By contrast, brighter particles reflect much incoming solar radiation back to space and thus have a net cooling effect. Further complexity in assessing the impact of dust results from the fact that dust aerosols have a relatively short life-time in the troposphere (a few hours to about a week) and show large variations in their temporal and spatial distribution (Hsu et al. 2000), both horizontally and vertically. Moreover, the radiative effects of a dust layer are modified by dynamical effects (e.g. convection) within the atmosphere (Harrison et al. 2001).

Because of this complexity, there is no clear consensus about whether substantially increased dust loadings at the Last Glacial Maximum (LGM) around 18 000–20 000 years ago could have caused additional cooling or could have caused warming (see, for example, Overpeck et al. 1996; Harrison et al. 2001; Claquin et al. 2003). In addition, it is possible that dust additions to ice caps and glaciers could modify their surface albedo, leading to changes in radiation budgets. Likewise, dust stimulation of phytoplanktonic production releases DMS which may increase cloud albedo and so contribute to cooling of the atmosphere (Henriksson et al. 2000).

3.10   Dust and Atmospheric CO_2

The presence of carbon dioxide in the atmosphere has been, is and will be a major influence on the radiation balance of the Earth. Carbon dioxide levels have varied through time and are believed to be one of the prime determinants of climate change. Dust loadings in the atmosphere may be interrelated with such changes. Ridgwell (2002), for example, has argued cogently that dust may affect climate by fertilizing ocean biota which in turn draw down CO_2 from the atmosphere, which in turn reduces the greenhouse effect. He believes that currently there are some parts of the ocean where a supply of Fe is a limiting factor in terms of phytoplankton growth. However, during the Ice Ages, when global dust production and deposition were considerably greater than today, it is possible that a series of feedbacks could lead to enhanced climatic change (Fig. 3.5). One scenario is that any intensification in glacial state would tend to produce an increase in dust availability and transport efficiency. This in turn could produce a decrease in CO_2 (through Fe fertilization of the Southern Ocean), which would cause further intensification in the glacial state and thus enhanced dust supply, and so one. As he argued (Ridgwell (2002, p. 2922):

"Operation of this feedback loop would come to an end once the global carbon cycle has reached a second state, one in which biological productivity becomes insensitive to further increases in aeolian Fe supply, perhaps through the onset of limitation by NO_3. If aeolian Fe supply were then to decrease sufficiently to start limiting biological productivity again, the feedback loop operating in the opposite direction would act so as to reverse the original climatic change. That the Earth system might exhibit two distinct states, one of 'high-xCO_2 low-dust' and the other 'low-xCO_2 high-dust', is consistent with developing views of the climate system as being characterized by the presence of different quasi-steady-states with abrupt transitions between them".

It is also possible, though as yet largely unproven, that dust may have encouraged growth of iron-hungry N_2-fixing cyanobacteria such as *Trichodesmium*, thus alleviating nitrate limitations (Pedersen and Bertrand 2000). In contrast, Maher and Dennis (2001) and Röthlisberger et al. (2004) suggested that the evidence for dust-mediated control of glacial–interglacial changes in atmospheric CO_2 is weak. They argue that dust peaks and CO_2 levels in the Vostok and Dome C ice cores show a mismatch and that, even in

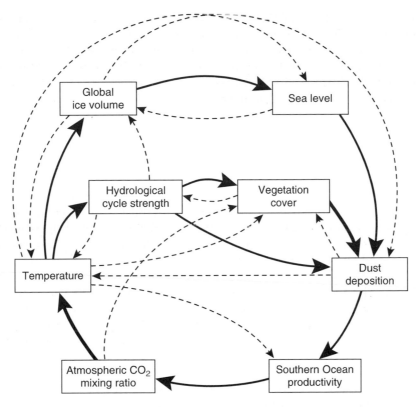

Fig. 3.5. Schematic diagram of the hypothetical glacial dust–CO_2–climate feedback system. Different components of the Earth system can directly interact in three possible ways: a positive influence (whereby an *increase* in one component directly results in an *increase* in a second – indicated by *red arrows* in the diagram), a negative influence (an *increase* in one component directly results in a *decrease* in a second – *black arrows*), or no influence at all. An even number (including zero) of negative influences occurring within any given closed loop gives rise to a positive feedback, the operation of which will act to amplify an initial perturbation. For instance, the two-way interaction apparent between temperature and ice volume is the 'ice–albedo' feedback. Conversely, an odd number of negative influences gives rise to a negative feedback, which will tend to dampen any perturbation. Primary interactions in the dust–CO_2–climate subcycle are indicated by *thick solid lines*, while additional interactions (peripheral to the discussion here) are shown *dotted* for clarity. Four main (positive) dust–CO_2-climate feedback loops exist in this system. 1. Dust supply → productivity → xCO_2 → temperature → ice volume → sea level → dust supply (four negative interactions). 2. Dust supply → productivity → xCO_2 → temperature → hydrological cycle → vegetation → dust supply (two negative interactions). 3. Dust supply → productivity → xCO_2 → temperature → hydrological cycle → dust supply (two negative interactions). 4. Dust supply → productivity → xCO_2 → temperature → ice volume → dust supply (two negative interactions). Modified after Ridgwell (2002, Fig. 11)

glacial periods, the dust flux supplied to the Southern Ocean was modest. Ridgwell and Watson (2002) believed this argument was overstated.

This 'iron hypothesis', first advanced by Martin et al. (1991), is the subject of considerable ongoing research (see, for example, Ridgwell 2003; Fan et al.

2004; Gao et al. 2003a); and the extent to which dust-stimulated phytoplankton growth leads to CO_2 drawdown of the magnitude shown in ice cores is still an uncertainty, though changes in the relative contribution of phytoplankton to total productivity during glacial cycles have been established through analysis of Tasman Sea cores by Calvo et al. (2004). Bopp et al.'s (2003) model indicated that the maximum impact of high dust deposition on atmospheric CO_2 must be less than 30 ppm.

3.11   Dust and Tropospheric Ozone

Another important way in which desert dust particles can affect the atmosphere is through their role in the photochemical production of ozone in the troposphere. Ozone concentrations have a whole suite of implications for humans and for other organisms.

Mineral dust appears to reduce the photolysis rates for ozone production by as much as 50% and provides reaction sites for ozone and nitrogen molecules. When being transported through the atmosphere, dust is frequently associated with nitrate and sulphate, the concentrations of which can increase with transport time (Savoie and Prospero 1982). This increase has been interpreted as implying that mineral aerosols may provide a reactive surface that is able to support heterogeneous processing of trace gases (Arimoto 2001). The measurement of ozone concentrations in dust plumes has confirmed these thoughts. Analysis over the Apennines in Italy showed that the lowest concentrations of ozone occurred during Saharan dust events (Bonasoni et al. 2004). In this study, the lowest ozone concentrations were recorded when the Saharan air masses were rich in coarse particles.

3.12   Dust and Clouds

Dust nuclei may modify cloud characteristics (Levin et al. 1996; Sassen et al. 2003). As Toon (2003, pp. 623–624) explained: "Dust may affect clouds in two ways. All water droplets start off by forming on pre-existing particles. As the number of particles increases, for instance due to a dust storm, the number of cloud droplets may increase. If there are more cloud droplets, the droplets will be smaller because the mass of condensing water is usually fixed by air motions and ambient humidity. Smaller cloud droplets make for a greater surface area and hence brighter clouds . . . A less well-studied phenomenon is that smaller droplets are also much less likely to collide with each other and create precipitation . . . By acting as nuclei for triggering ice formation, dust particles can also affect clouds by causing the water droplets to freeze at higher temperatures than expected . . . Dust may thus be triggering precipitation in low-altitude clouds that otherwise would be too warm to have produced rain, or be triggering rain at lower levels in convective clouds that otherwise would not have produced rain until reaching much higher altitudes where it is colder ... Dust may therefore inhibit precipitation by making more and smaller droplets, or enhance it by adding ice particles to warm clouds".

Rosenfeld et al. (2001) argued that the inhibiting effect on precipitation was most likely and that Saharan dust provides very large concentrations of cloud condensation nuclei, mostly in the small size range, which mean that clouds are dominated by small droplets so that there is little coalescence. This results in suppressed precipitation, drought enhancement and more dust emissions, thereby providing a possible desertification feedback loop.

Desert dust is also undoubtedly associated with strong ice-nucleating behaviour (Sassen et al. 2003; Sassen 2005) and high concentrations of dust particles acting as ice nuclei in clouds could lead to changes in cloud microphysical and radiative properties, latent heating and precipitation. Interest has started to build in recent years in the possible role that Saharan dust plays in modifying convective storm activity – anvil cloud development and precipitation – over Florida (Van Den Heever et al. 2005).

Another way in which rainfall may be affected is through changes in convective activity brought on by the modification of temperature gradients in the atmosphere created by the presence of dust (Maley 1982). In addition, the radiative effects of dust may lead to the intensification of easterly waves in North Africa (Jones et al. 2003) with consequent effects on numerous climatic parameters, including precipitation. One study of outbreaks of dust-laden Saharan air over the Atlantic – the so-called Saharan air layer, or SAL – suggests that they may inhibit the intensification of tropical waves, tropical disturbances, or pre-existing tropical cyclones due to the SAL's dry air, temperature inversion and strong vertical wind shear associated with the mid-level easterly jet (Dunion and Velden 2004). They may suppress convection (Wong and Desler 2005). It is probable that dust loadings in the atmosphere were both affected by past climatic changes and had an effect on such changes through complex feedback processes (Harrison et al. 2001).

3.13 Economic Effects

The entrainment, transport and deposition of dust can present a variety of problems to inhabitants in and around desert areas (see Tables 1.1 and 3.1), many of which have a deleterious economic impact. Such hazards have affected dryland peoples since time immemorial. Folk (1975), for example, suggests that the ancient Macedonian town of Stobi, which flourished between 400 BC and 400 AD, was abandoned because of the severe affects of dust storms.

A more recent example of the mix of impacts a dust storm can bring is provided for China by Yang et al. (2001, p. 49):

"A major sand-storm on May 5th 1993 caused serious economic loss and was as hazardous as a disaster caused by an earthquake. According to ground observation and investigation made by the expert group of the Ministry of Forestry, a total of 85 people died, 31 people were lost and 264 were injured (most of these victims were children). Agriculture and animal husbandry were most severely hurt. In total, 373,000 ha of crops were destroyed. 16,300 ha of fruit trees were damaged. Thousands of greenhouses and plastic mulching sheds were broken. 120,000 heads of animals died or were irrecoverably lost. The fundamental agricultural installations and grassland service facilities were ruined. More than 1,000km of irrigation channels was buried by sand accumulation. Many water resource back-up facilities, such as reservoirs, dams, catchments, underground canals and flood control installations were filled up with sand silts. About 6,021 communication poles and electricity grids were pushed down and electricity transports and communication services in some regions were stopped for several days. Some sections of railway and highway were interrupted due to deflation and sand accumulation."

Another major dust and sandstorm event took place in April 2002 and led to airport closures in Mongolia and Korea. The total damage cost of this event in Korea alone was put at U.S.$ 4.6 billion (or about 0.8% of GDP; Asian Development Bank 2005, pp. 1–5).

In a similar vein, dust storms have regularly been associated with deaths in India. In April 2005, ten people and 50 head of cattle were killed by fires fanned by dust storm winds in Uttar Pradesh. In March 2005, six people were killed and 40 injured in a dust storm in Bihar.

Some progress has been made in identifying the offsite costs of wind erosion. In South Australia, for example, the costs include damage to houses and the need for redecoration, the need to clean power transformers, deaths and damage caused in traffic accidents, road disruption, impacts on the costs of air travel and impacts on human health (especially because of raised asthma incidence – see Section 3.14 below; Williams and Young 1999).

The reduction in visibility caused by dust storms is a hazard to aviation, rail and road transport (Fig. 3.6). The severe pre-frontal storm of 7 November 1988 in South Australia, for example, caused road and airport closures all across the Eyre Peninsula (Crooks and Cowan 1993). In the United States, in November 1991, a series of collisions involving 164 vehicles occurred on Interstate 5 in the San Joaquin Valley in California (Pauley et al. 1996), while in Oregon a dust storm in September 1999 set off a chain reaction of 50 car crashes that killed eight people and injured more than 20 (State of Oregon 2004). The loss of visibility may be very sudden when caused by the arrival of a dust wall associated with a dry thunderstorm. Such Haboob dust walls were responsible for 32 multiple accidents between 1968 and 1975 on Interstate 10 in Arizona (Brazel and Hsu 1981). The seriousness of the problem inspired the development of a Dust Storm Alert System involving remote-controlled

Fig. 3.6. Dust and sand storms pose considerable problems for transport links, here the block-ing of the main railway line between Walvis Bay and Swakopmund in Namibia (from ASG)

road signs and special dust-alert messages broadcast on local radio (Burritt and Hyers 1981).

Some fatal commercial air crashes have also been attributed to visibility reduction or to the adverse mechanical effects of dust storms. On 7 May 2002, for example, an EgyptAir aircraft crashed near Tunis, killing 18 of 60 people on board. On 30 January 2000, a Kenya Airways Airbus crashed in the Ivory Coast with the loss of 179 lives.

3.14   Health

A number of medical conditions can be traced to the impact of desert dust; and the effects of fine wind-borne particles on human health have recently been the subject of considerable interest (Griffin et al 2001; Garrison et al. 2003). On 9 August 2005, a dust storm in Baghdad led to nearly 1000 cases of suffocation being reported to the city's Yarmuk Hospital, one of whom died. The straightforward inhalation of fine particles can cause and/or aggravate diseases such as bronchitis, emphysema and silicosis. High incidences of silicosis and pneumoconiosis have been reported in Bedouins in the Negev (Bar-Ziv and Goldberg 1974), while dust blown by the *Irifi* wind of the Western (formerly Spanish) Sahara is responsible for the conjunctivitis that is common among the nomads of the country (Morales 1946). High concentrations of atmospheric dust in many desert areas often exceed generally recommended health levels for particulate matter (see also Section 6.1 on PM_{10} values). In Mali, for example, Nickling and Gillies (1993) found that the mean ambient air concentrations during April–June were 1176 μg m^{-3}, exceeding the recommended international health standard by an order of magnitude. Similar concentrations can also occur during particularly severe long-range transport events. In certain parts of Spain, the levels of particulate matter associated with frequent incursions of dust from North Africa means that it is not possible to meet European Union directives on acceptable levels of air pollution (Querol et al. 2004). Rodriguez et al. (2001) indicated that these Saharan dust events can induce up to 20 days a year in which PM_{10} standards are exceeded in southern and eastern Spain. Intrusions of desert dust from the Hexi Corridor in northern China also make a significant contribution to particulate pollutants in the Lanzhou Valley, an urban area that is among the worst in China for its poor air quality (Ta et al. 2004b).

Dust may also be otherwise contaminated by organisms, such as bacteria and fungi (Kellogg et al. 2004), and by toxic chemicals that can harm people when it settles on the skin, is swallowed or inhaled into respiratory passages. The increase in dust storm activity in Turkmenistan, for example, linked to the desiccation of the Aral Sea, has probably caused severe respiratory problems for children in the area, but the dust from the dry sea bed also happens to contain appreciable quantities of organophosphate particles (O'Hara et al. 2000). Dust blown from another former lake bed, that of the desiccated Owens Lake in California, contains arsenic derived from nineteenth-century mining operations (Raloff 2001). Dust storm material in Saudi Arabia has been found to contain an array of aeroallergens and antigens which could trigger a range of respiratory ailments (Kwaasi et al. 1998). Other possible consequences of airborne dust include an increase in asthma incidence (Rutherford et al. 1999), as reported for Barbados and Trinidad when Saharan dust outbreaks occur (Monteil 2002; Gyan et al. 2005), and also an increase in the incidence of meningococcal meningitis in the Sahel zone and Horn of Africa (Molesworth et al. 2002). The annual meningitis epidemics in West Africa, which affect up to 200 000 people between February and May, are closely related to the Harmattan season in their timing (Sultan et al. 2005).

Coccidioidomycosis, a disease caused by a soil-based fungus (*Coccidioides immitis*) transported in airborne dust, is endemic to parts of the southwestern United States (especially in the San Joaquin Valley of California, southern Arizona, southern New Mexico and west Texas) and northern Mexico (Gabriel et al 1999). In the United States, where it is known as Valley Fever, an estimated 50 000–100 000 persons develop symptoms of the disease each year (Leathers 1981); and a dramatic increase in the incidence of coccidioidomycosis during the early 1990s in California was estimated to have cost more than U.S.\$66 million in direct medical expenses and time lost in one county alone (Kirkland and Fierer 1996).

Dust can also contain dried rodent droppings or urine which can cause the spread of Hantavirus Pulmonary Syndrome. In Ladakh and China, dust may contribute to a high silicosis incidence (Derbyshire 2001); and fungal spores from China reach high ambient levels in Taiwan during dust events and may have health implications (Wu et al. 2004). Some recent epidemiological studies indicate that long-range dust transport events are closely associated with an increase of daily mortality in Seoul, Korea (Kwon et al. 2002), and Taipei, Taiwan (Chen et al. 2004), and caused cardiovascular and respiratory problems (Kwon et al. 2002), including an increased incidence of strokes (Yang et al. 2005).

Given the great distances over which dust can be transported, it is not surprising to learn that the intercontinental dispersal of material may include pathogens of crop plants. Long-distance dispersal of fungal spores by the wind can spread plant diseases across and between continents and re-establish diseases in areas where host plants are seasonally absent (Brown and Hovmøller 2002). While monitoring aerosols on the Caribbean island of Barbados, Prospero (2004) reported that concurrent detection of bacteria and fungi only occurred in air that contained Saharan dust.

3.15 Dust Storms in War

Large-scale military movements in desert environments can be both the cause and the victim of dust events. The disruption of desert surfaces during the North African campaign in the 1940s increased the occurrence of dust storms in the region to a considerable extent (Oliver 1945). The significance of dust storms for military activities again became apparent during the Gulf War of 1990–1991 and the Iraqi War of 2003–2004. In April 2005, 18 people were killed when a United States military Chinook helicopter came down in a heavy dust storm in Ghazni, Afghanistan.

The human implications of dust storms were graphically illustrated during the North African campaign. In the summer of 1941, Titch Cave, member of a Long Range Desert Group (LRDG) patrol that had just come in from the desert, witnessed a storm at Siwa oasis in Egypt just as he and his colleagues were to sit down and have a rare meal of fresh meat (Morgan 2000, p. 85):

"The mutton was carefully cooked, while we all waited in anticipation, and after being carved was just ready to be served when an excited voice from outside shouted, '**** me! Come and look at this.'

"We all dashed out not knowing quite what to expect and there, all across the northern horizon, was a huge rolling cloud which must have been over 100 feet high. We watched in awe, our dinner forgotten, as the cloud rolled down over the northern cliffs and advanced towards us across the oasis. The air was quite still as the cloud approached, then, when it was closer, the wind began to rise, the temperature dropped and it was upon us, filling the air and every nook and cranny of our hut with dust and sand.

"It was the father and mother of a sandstorm which was beyond the experience of even the oldest members of our patrol. Of course, our dinner was ruined . . ." Field Marshal Rommel also wrote graphically about a storm, locally called the *Ghibli*, which took place in Libya in March 1941, an account that also reinforces the dust hazard to transport (Liddell Hart 1953, p. 105):

"After taking off . . . we ran into sandstorms near Taourga, whereat the pilot, ignoring my abuse and attempts to get him to fly on, turned back, compelling me to continue the journey by car from the airfield at Misurata. Now I realized what little idea we had of the tremendous force of such a storm. Immense clouds of reddish dust obscured all visibility and forced the car's speed down to a crawl. Often the wind was so strong that it was impossible to drive along the Via Balbia. Sand streamed down the windscreen like water. We gasped in breath painfully through handkerchiefs held over our faces and sweat poured off our bodies in the unbearable heat. So this was the Ghibli. Silently I breathed my apologies to the pilot. A Luftwaffe officer crashed in a sandstorm that day."

Sandstorms are not only uncomfortable for the military personnel forced to endure them. They can also be damaging to their vehicles and armaments as well. This was well described by one of the soldiers in Popski's Private Army, a special unit that operated behind enemy lines in the Second World War. As Park Yunnie wrote (Yunnie 2002, p. 20):

"It hit us like a whip-lash, taking our breath, leaving us cowed and defenceless, whimpering with pain. We couldn't breathe. Hot, smarting dust clogged our nostrils, seared the backs of our throats, coated our tongue and gritted in our teeth; drifts of fine-blown sand formed in the folds of our clothing, blew into our pockets and found its way through to our skins; sand piled up in the trucks, forming miniature dunes, stuck to the oily and greasy parts of the chassis, blew under the bonnet and sifted into the carburetor, the magneto, the unsealed working parts; grating sand filtered into the Vickers guns, jamming the ammunition pans; sand found its way into everything, everywhere. Each truck was isolated in its own drift, cut off from the others by an impenetrable wall of frenzied shrieking grit . . ."

The side of his truck was polished like a mirror, every vestige of paint sanded off.

4   The Global Picture

4.1   Introduction

The fact that dry, unprotected sediments can be entrained by wind in almost any physical environment is reflected in the large number of names in common use for dust-bearing winds (Table 4.1). Nonetheless, the major source regions of contemporary mineral dust production are found in the desert regions of the northern hemisphere, in the broad swathe of arid territory that stretches from West Africa to Central Asia, while lesser sources are found in the world's other major desert areas. This global picture of desert dust production has been pieced together using satellite imagery and standard terrestrial meteorological observation data, but the details are still not complete.

Satellites represent the only data source with truly global coverage and analysis of their data has produced some of the best global surveys of dust storm distribution. The Total ozone mapping spectrometer (TOMS) has proved to be among the most effective instruments for detecting atmospheric mineral dust (Herman et al. 1997; Prospero et al. 2002; Washington et al. 2003). We also have global or near-global maps of aerosol optical thickness (a measure of aerosol column concentration) derived from satellites such as the NOAA Advanced very high resolution radiometer (AVHRR) and MODIS (see, for example, Chin et al. 2004; Ginoux et al. 2004; Yu et al. 2003). Global images are available on http://www.osdpd.noaa.gov/PSB/EPS/Aerosol/Aerosol.html (accessed 22 June 2005).

4.2   Major Global Sources

TOMS data have been used to derive an Aerosol Index (AI), values for which are linearly proportional to the aerosol optical thickness. The world map of annual mean AI values (Fig. 4.1) has certain clear features. First, the largest area with high values is a zone that extends from the eastern subtropical Atlantic eastwards through the Sahara Desert to Arabia and southwest Asia. In addition, there is a large zone with high AI values in central Asia, centred over the Taklamakan Desert in the Tarim Basin. Central Australia has a relatively small zone, located in the Lake Eyre basin, while southern Africa has

Table 4.1. Dust-bearing winds. After Olbruck (1973), Goudie (1978), Nalivkin (1983), Middleton (1986c) and other sources

Region	Wind (location)
Asia	Afganets (Tadjikistan)
	Garmsil (Turkmenistan)
	Kara Buran (Central Asia)
	Ibe (Kazakhstan)
	Balkhash Bora (Kazakhstan)
	Loo (India)
	Andhi (India)
	Kyzyl Buran (China)
	Yaman (China)
	Hyi Fyn (China)
	Huan Fyn (China)
	Shachenbao (China)
	Fuhjin (Japan)
	Kosa (Japan)
	Huang Sa (Taiwan, Korea)
Middle East	See Table 5.8
Europe	Calina (Spain)
	Leveche (Spain)
	Kossava (Hungary)
	Scirocco (S. Europe)
	Sukhovey (S. Russian steppe)
	Chernye Buran (Russia and Ukraine)
	Blow (England)
	Mistur (Iceland)
Latin America	Chubasco (Mexico)
	Tolvanera (Mexico)
	Paracas (Peru)
	Pampero Sucio (Argentina)
	Volcan (Argentina)
	Zonda (Argentina)
N. America	Chinook (USA – Rocky Mountains)
	Keeler Fog (USA – California)
	Palouser (USA – Idaho, Montana)
	Santa Ana (USA – California)
	Wasatch (USA – Utah)
Australia	Bedouries (W. Queensland)
	Brickfielder (Victoria)
	Cobar Shower (New South Wales)
	Darling Shower (New South Wales)
E. and S. Africa	Kharif (Somalia)
	Gobar (Ethiopia)
	Berg Wind (Namibia)
Sahara	See Table 5.1

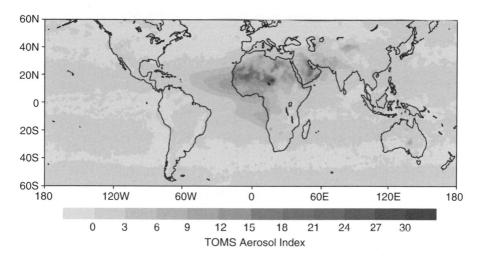

Fig. 4.1. The world map of annual mean aerosol index values determined by TOMS

two zones, one centered on the Mkgadikgadi basin in Botswana and the other on the Etosha Pan in Namibia. In Latin America, there is only one easily identifiable zone. This is in the Atacama and is in the vicinity of one of the great closed basins of the Altiplano – the Salar de Uyuni. North America has only one relatively small zone with high values, located in the Great Basin. Other satellite-derived maps of aerosol optical thickness show a generally very similar picture of dust loadings in the atmosphere.

The importance of these different dust 'hot spots' can be gauged by looking not only at their areal extents, but also at their relative TOMS AI values. Table 4.2 lists the latter. This again brings out the very clear dominance of the Sahara in particular and of the Old World deserts in general. The Southern Hemisphere as a whole and the Americas are both notable for their relatively low AI values. So, for example, the AI values of the Bodélé Depression of the south central Sahara are around four times greater than those recorded for either the Great Basin or the Salar de Uyuni. However, the best way to assess the relative importance of dust source areas on a global basis is to combine their areas and their AI values (Fig. 4.2). This again brings out the enormity of the Saharan dust source in comparison with those of Arabia, China and the Thar.

Thus, analysis of TOMS data enables a global picture of desert dust sources to be determined. It demonstrates the primacy of the Sahara and highlights the importance of some other parts of the world's drylands, including the Middle East, Taklamakan, southwest Asia, central Australia, the Etosha and Mkgadikgadi pars of southern Africa, the Salar de Uyuni in Bolivia and the Great Basin in the United States. One characteristic that emerges for most of these regions is the importance of large basins of internal drainage as dust sources (Bodélé, Taoudenni, Tarim, Seistan, Eyre, Etosha, Mkgadikgadi, Etosha,

Table 4.2. Maximum mean AI values for major global dust sources determined from TOMS

Location	AI value	Average annual rainfall (mm)
Bodélé Depression of south central Sahara	>30	17
West Sahara in Mali and Mauritania	>24	5–100
Arabia (Southern Oman/Saudi border)	>21	<100
Eastern Sahara (Libya)	>15	22
Southwest Asia (Makran coast)	>12	98
Taklamakan/Tarim Basin	>11	<25
Etosha Pan (Namibia)	>11	435–530
Lake Eyre Basin (Australia)	>11	150–200
Mkgadikgadi Basin (Botswana)	>8	460
Salar de Uyuni (Bolivia)	>7	178
Great Basin of the USA	>5	400

Uyuni and the Great Salt Lake). Related to this is the fact that many sources are associated with deep and extensive alluvial deposits (Prospero et al. 2002). In contrast, sand dune systems are not good sources of fine-grained dust.

Dust storms also occur under cold climate conditions. They have been described, for example, from outwash plains in Iceland (Fig. 4.3), deltas in Alaska, sandurs in Baffin Island and braided river beds in New Zealand (Seppälä 2004).

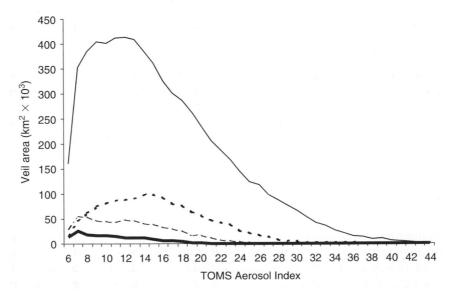

Fig. 4.2. Area average TOMS aerosol index values for the main dust regions: Sahara (*solid line*), Arabia (*heavy dashed line*), Thar (*light dashed line*), northwest China (*heavy solid line*)

Fig. 4.3. Dust blowing into the North Atlantic from southern Iceland, 28 January 2002 (MODIS)

Estimates of the total soil dust emissions to the atmosphere on a global scale (Table 4.3) show a large range (see the excellent review by Prospero 1996a), largely because models vary with regard to such factors as the rate of scavenging of particles from the air. A discussion of the relative contributions made by the Sahara and other major sources can be found in Section 5.5.

4.3   Dust Storms and Rainfall

Because rainfall amounts affect two important controls of dust storm activity – soil moisture and vegetation cover – it is to be expected that dust storm occurrence will broadly be inversely correlated with rainfall amount. Plainly, very wet areas removed from dust source areas by some distance do not have many dust storms (Goudie 1983). Indeed, Goudie (1983), on the basis of analysis of terrestrially observed meteorological data, argued that dust storm

Table 4.3. Estimates of dust emissions to the global atmosphere

Author (s)	Annual quantity ($\times 10^6$ t)	Atmospheric turnover time (days)
Peterson and Junge (1971)	500	
Schütz (1980)	<5000	
D'Almeida (1986)	1800–2000	
Tegen and Fung (1994)	3000	
Andreae (1995)	1500	4.0
Duce (1995)	1000–2000	
Mahowald et al. (1999)	3000	5.1
Luo et al. (2003)	1654	5.1
Zender et al. (2003)	1490	4.3
Ginoux et al. (2004)	1950–2400	7.1
Liao et al. (2004)	1784	3.9
Miller et al. (2004b)	1019	5.2

frequency is at a peak in areas where the rainfall is between 100 mm and 200 mm per annum and that in areas with rainfall <100 mm, dust storm frequency appears to decline. He advanced three possible explanations for this decline: (1) such areas may have smaller amounts of alluvium supplied by streams which could act as a dust source, (2) in very dry areas there is insufficient exposure to dust-producing atmospheric conditions associated with convective activity or the passage of fronts and (3) the reservoir of available source material may have been depleted by long-continued wind activity and by the formation of some wind-stable surfaces, such as stone pavements. Additionally, such areas may suffer from less human disturbance than more humid areas, which tend to be heavily grazed and farmed.

However, TOMS data indicate that many of the world's major dust source regions are areas of hyperaridity. The prime global source, the Bodélé Depression, has a mean annual rainfall of some 17 mm (at Faya Largeau), while the large West Sahara source has annual precipitation levels between 5 mm and 100 mm. In Arabia, dust storms are most prevalent where the mean annual rainfall is <100 mm (Goudie and Middleton 2001); and the great Taklamakan dust source in north-west China has large areas where the rainfall totals are <25 mm per annum.

Two coherent explanations of this contradiction – between Goudie's conclusions and those based on the TOMS data – involve appraisals of the sources of data used. The terrestrial meteorological stations on which Goudie (1983) based the relationship between dust-raising and annual rainfall are relatively sparse in many of the driest desert regions (Fig. 4.4) and thus the relationship may be more apparent than real. However, this is not to say that TOMS is a perfect data source. The TOMS AI is known to be sensitive to the

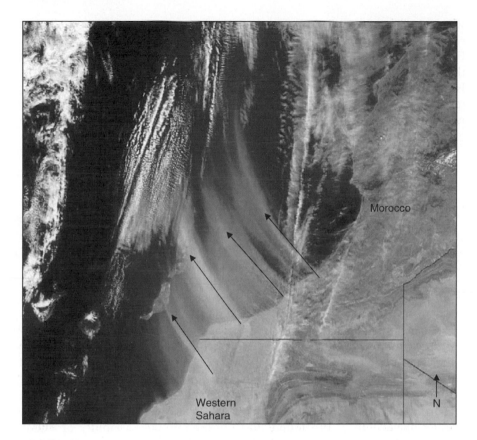

Fig. 4.4. Curved dust plumes emanating from southern Morocco and northern Western Sahara, an area with very few terrestrial meteorological observing stations, on 12 March 2003 (MODIS)

height of the aerosol layer (Torres et al. 1998) and may underestimate the importance of further dust sources on the edge of deserts, areas where boundary layer heights are lower (Mahowald and Dufresne 2004).

For North Africa, location of the largest global sources, this quandary is not new. Prospero (1996b) raised the question of 'Sahara vs Sahel' as the primary source for transatlantic dust transport. The fact that dust concentrations at Barbados are inversely related to the previous year's rainfall in Sahelian Africa suggests a link with sources in the Sahel. Indeed, several studies have shown the importance of areas in Sahelian latitudes as sources that have increased their dust output during recent periods characterized both by prolonged drought and intensified land use, in places leading to desertification (Middleton 1985a, b; Goudie and Middleton 1992; N'Tchayi et al. 1997).

Indeed, generally the link between drought phases and enhanced dust production is well established, though there will be different time-lags in different areas (Zender and Kwon 2005). This relationship is close in Australia, where peaks in annual dust storm frequency seen in the meteorological records are clearly related to drought phases (McTainsh et al. 2005). However, the relationship becomes less certain in the light of the character of the Sahara's two main dust source areas as highlighted by TOMS. While Harmattan dust blown from the Bodélé Depression tends not to travel far over the Gulf of Guinea, dust entrained in the West Sahara source area does make a significant contribution to transatlantic flows. However, while the effect of drought on dust-raising in Sahelian latitudes can be established, reference to 'drought' in an area receiving less than 100 mm in mean annual rainfall is less sound.

Two explanations for the established increase in trans-Atlantic dust transport in recent decades can be proposed. Firstly, that flows from a constant West Sahara source area have been augmented by material from more southerly, drought-affected sources. Secondly, it is possible that the relationship between Barbados dust concentrations and Sahelian rainfall reflect other changes in atmospheric circulation associated with drought in the Sahel. As Prospero (1996b) has pointed out, both the Hadley circulation and the mid-tropospheric easterly jet are more intense during Sahelian dry spells (Nicholson 1986; Newell and Kidson 1984). The second hypothesis is not necessarily incompatible with the first. Stronger winds, be they low-level entraining airflows, upper tropospheric transporting flows, or both, could mean more material being transported from the West Sahara source area and/or from Sahelian source areas. Indeed, work in Australia indicates that drought periods may be associated with higher overall wind speeds (McTainsh et al. 1999).

Part of the conundrum in North Africa may have been explained by Moulin and Chiapello (2003) who found that the interannual variability of TOMS atmospheric dust optical thickness over the Atlantic in the summer months (June–August) was very largely controlled by dust emissions in the northern Sahel (15–17° N). These authors also established a large-scale correlation between summertime Atlantic dust export and the occurrence of drought in the Sahel, suggesting that the variability of Sahelian dust emissions are primarily controlled by the position of the vegetation boundary with the Sahara, a border that other satellite-based studies have demonstrated is highly dynamic from one year to the next (Tucker and Nicholson 1999).

4.4   Vegetation and Dry Lake Beds

At the global scale, Engelstädter (2001) analysed the importance of dry lake beds (Fig. 4.5) and vegetation type in controlling the occurrence and frequency of dust storms. Vegetation types were derived from the BIOME 4 model and from the satellite-derived NDVI (Normalized difference vegetation index). The extent of palaeolake beds was based on the surface hydrological transport model HYDRA (Hydrological routing algorithm).

Fig. 4.5. A small dust storm being generated from the dry floor of an old lake bed in the Wadi Rum area of southern Jordan (from ASG)

The results of his analyses of dust storm frequency in relation to vegetation types are shown in Tables 4.4 and 4.5. As might be anticipated there is a general tendency for areas of limited vegetation cover to be associated with high dust storm incidence. The results of his analyses of the effect of dry palaeolake beds are shown in Table 4.6. Areas with a high percentage covered by such depressions have markedly higher dust storm frequencies than those that do not.

4.5   Diurnal and Seasonal Timing of Dust Storms

Various observers have established that dust storms tend to be concentrated in certain parts of the day. For example, in Turkmenistan, Orlovsky and Orlovsky (2001) found that the number of dust storms was greatest in the late morning and afternoon (Fig. 4.6) and was caused by a wind speed maximum at that time and drying of the soil in the daylight hours. Similarly in the Gobi of Mongolia (Fig. 4.7), most dust storms occur in the afternoon and relatively few by night (Middleton 1991; Natsagdorj et al. 2003). Wang et al. (2005a) identified a similar pattern in China. Tsunematsu et al. (2005), working on the Taklamakan Desert, suggested that many dust outbreaks occurred following the breakdown of a nocturnal inversion (which developed over the basin at night and promoted atmospheric stability) during the late morning.

Table 4.4. Dust storm frequency in relation to vegetation types and net primary productivity (NPP) modelled from BIOME4. From Engelstädter (2001, Section 4.2.2., Table 1)

Biome type	Number stations	Average (median) of dust storm frequency (days year^{-1})	Average annual NPP (g C m^{-2} year^{-2})
Desert and barren	96	5.7	119.5
Temperate xerophytic shrubland	95	2.1	292.0
Temperate evergreen needleleaf open woodland	11	0.4	299.0
Grassland	47	2.0	466.0
Tropical xerophytic shrubland	57	1.4	424.0
Temperate schlerophyll woodland and shrubland	35	0.3	576.0
Forest	95	0.2	828.0
Tropical savanna	10	0.1	904.5
Temperate deciduous broadleaf savanna	6	0.2	1059.5

In Belarus, the majority of dust storms take place before 1500 hours (Chizhikov and Kamlyuk 1997).

In the Middle East, Middleton (1986a) established that much of the area witnessed dust storm maxima during daylight when intense solar heating of the ground surface creates a high degree of turbulence and very strong pressure gradients locally. This is shown dramatically for Kuwait, where about 50% of the dust storm hours occur between about 1200 hours and 1700 hours (Safar 1985; Fig. 4.8).

Table 4.5. Average annual dust storm frequency and number of stations for different vegetation types derived from the DeFries data set. From Engelstädter (2001, Section 4.2.2., Table 2)

Vegetation type	Number of stations	Average (median) annual dust storm frequency (days year^{-1})
Bare ground	71	6.8
Shrubs and bare ground	96	2.6
Grassland	181	1.6
Cultivated crops	140	0.8
Wooded grassland	69	0.1
Broadleaf deciduous forest and woodland	23	0.3
Forest	28	0.1

Table 4.6. Statistical data of potential dry lake bed fractions for: (a) areas of *desert and barren, temperate xerophytic shrubland, grassland* and *tropical xerophytic shrubland* and (b) areas of *desert and barren.* From Engelstädter (2001)

Dry palaeo-lake bed fraction (%)	Number of stations	Average (median) dust storm frequency (days year⁻¹)
Group (a) areas		
0–20	222	2.3
20–40	36	2.2
40–60	14	6.1
60–80	5	14.9
80–100	10	24.5
Group (b) areas		
0–20	61	3.5
20–40	14	6.2
40–60	6	14.0
60–80	5	14.9
80–100	10	24.5

In the United States, Orgill and Sehmel (1976) found that the afternoon maximum in dust frequency is common and occurs when the atmosphere boundary layer is normally deep and turbulent mixing is more pronounced. This was confirmed for North Dakota by Godon and Todhunter (1998) who found that nearly 70% of dust events occurred between 1200 hours and 1800 hours. Stout (2001), working in the High Plains, recognized that there was

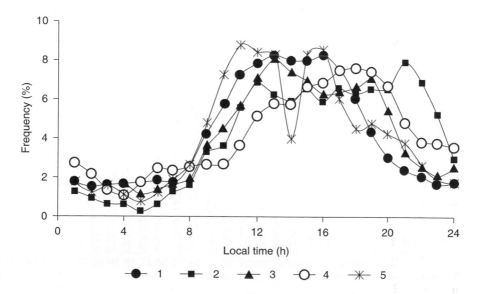

Fig. 4.6. Diurnal pattern of dust storms in Turkmenistan for five stations during 1981–1995. Modified after Orlovsky and Orlovsky (2001, Fig. 3)

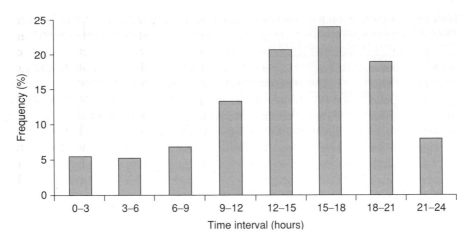

Fig. 4.7. The daily variation of dust storms for spring in the Gobi. Modified after Natsagdorj et al. (2003, Fig. 6)

typically a daily cycle of relative humidity driven by daily temperature variations with the lowest relative humidities occurring around mid-afternoon. This would be when surface soil moisture would be least and the potential for blowing dust at its maximum.

In the Saharan source regions, the dust maximum occurs between late morning and mid-afternoon (N'Tchayi et al. 1997).

The seasonality of dust storms is affected by a number of factors. These include: rainfall regimes (which control soil moisture conditions), seasonal

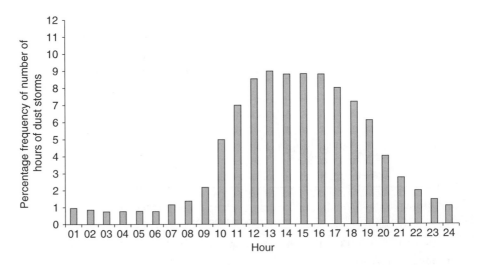

Fig. 4.8. Diurnal variation of dust storms at Kuwait International Airport (1962–1982). Modified after Safar (1985, Fig. 18)

snow cover (which may prevent soil deflation in winter months), the desiccation of closed lake basins, wind energy conditions, convectional activity and the passage of dust-raising depressions, and agronomic practices (which leave the soil bare in certain seasons). Littmann (1991a) attempts to categorize patterns of dust storm seasonality and shows, for example, the dominance of spring dust storms in China, the prevalence of pre-south westerly monsoon dust storms in India and the summer (dry season) maximum in the Middle East. Areas with a Mediterranean rainfall regime tend to have limited dust activity in the wet, winter months, whereas tropical regions with a strong summer rainfall regime have limited dust storm activity at the height of the wet season. Details of seasonality are given for each major region in Chapter 5, and are summarized in Table 4.7.

Table 4.7. Seasons or months of maximum dust storm activity

Location	Season (months)
Argentina	Winter (April–Aug.)
Arizona	Summer
Bahrain	Summer (March–July)
Belarus	Spring (April–May)
Bodélé	Early summer (April–June)
Canada	Spring
Chile	Winter (May–Sept.)
Egypt	Winter/Spring (Dec.–May)
Etosha (Namibia)	Autumn (Aug.–Nov.)
Eritrea	Summer (June–Aug.)
India/Pakistan	Early summer (May–June)
Kazakhstan	Summer (May–Aug.)
Korea	Spring (March–May)
Mexico City	Spring (Jan.–May)
Mongolia	Spring (April–May)
Northern Sahara	Summer (April–Aug.)
Queensland, northern New South Wales	Spring/early Summer
Sahel	Winter (Nov.–March)
Seistan basin (Afghanistan)	Summer (May–Sept.)
Taklamakan	Late Winter/Spring
Tokar delta (Sudan)	Summer (June–July)
Turkmenistan	Spring
United States of America	Spring
Victoria, southern New South Wales	Summer

4.6   Duration of Dust Storms

Dust storms do not generally last for very long. In Turkmenistan, for example, the frequency of dust storms with a duration of 12 h or more is only about 3%, though occasional examples lasting three days have been observed there (Orlovsky and Orlovsky 2001). In the Gobi of Mongolia, the average dust storm lasts 1.6–6.0 h (Natsagdorj et al. 2003), while in China's Taklamakan, the most serious dust storm conditions persist for 2–4 h (Yoshino 1992). More generally in China, Wang et al. (2005a) suggest that most dust storms last <2 h.

4.7   Dust Storms on Mars

Although this book is about dust storms on Earth, aeolian processes are active on other planets in the solar system (Greeley and Iversen 1985) and the transport and deposition of dust is particularly important on Mars, which has become the focus for a growing body of research. The 'Red Planet' could just as easily be called the 'Dust Planet', for yellowish brown dust gives the planet its colour. Dust storms occur almost daily, with thousands occurring each year (Cantor 2003). Telescopic observations since the eighteenth century and images delivered by spacecraft missions have shown Mars to be an arid planet dominated by the presence of dust both suspended in the atmosphere and deposited widely over the planet's surface. Features like meso-scale linear streaks, up to 400 km in length, are indicators of the power of dust entrainment (Thomas et al. 2003).

   Dust events on Mars have been observed at all scales, ranging from local dust devils (Balme et al. 2003) to storms that envelop the entire planet, dubbed Mars global dust storms, or GDSs (Fig. 4.9). These planet-encircling dust storms occur approximately one year in three, usually in late southern spring, (i.e. near perihelion) when Mars is closest to the Sun (Zurek and Martin 1993; Pankine and Ingersoll 2004). During the Martian summer, in the lower boundary layer of its clear, thin, cold atmosphere, the large temperature gradient that exists above the relatively warm surface may support intense free convection and the formation of dust devils. These can be greater than those found on Earth, commensurate with the deeper convective boundary layer on Mars during summer, reaching several hundred metres across and 8 km high. Regional dust storms may be produced whenever the poleward temperature gradient is sufficiently large to generate intense zonal circulation across the mid-latitudes in the form of baroclinic waves. Other regional dust storms are produced by katabatic outflow from receding frost outliers of the polar caps, with winds descending from areas of high relief (Fig. 4.10). Regional dust storms affect the radiation budget and this can lead to feedback effects that may cause the development of dust storms of global

Fig. 4.9. The surface of Mars on a relatively clear day (*left*, 26 June 2001) in contrast to a global dust storm (*right*, 4 September 2001). Both images from the Hubble space telescope

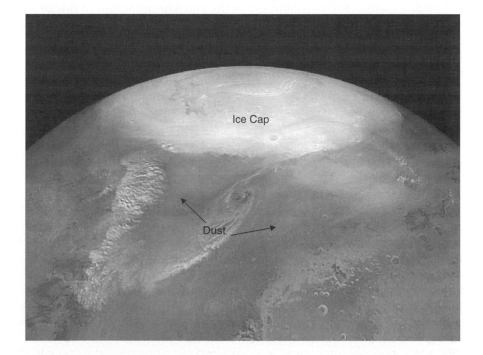

Fig. 4.10. Regional dust storms over Mars on the margin of the great polar ice cap

dimensions (Wells and Zimbelman 1997). Such an event was detected by the Hubble Space telescope in June 2001 (Strausberg et al. 2005, Plate 4.4). What was to become the biggest event for about a quarter of a century began as a small dust cloud inside the Hellas Basin (a deep impact crater in Mars's southern hemisphere). By early July, the dust cloud had spilled out of the basin and engulfed the whole planet. It is possible that airborne dust particles absorb sunlight and warm the Martian atmosphere strongly in their vicinity. Warm pockets of air spread quickly towards colder regions, thereby generating strong winds. These lift more dust off the ground and so create a positive feedback. In this model, dust heating seems to play a broadly analogous role to the release of latent heat in moist convection during the development of tropical storms and hurricanes on Earth (Read and Lewis 2004, p. 207).

It seems that dust has long played a fundamental role in the Martian climate (Greeley et al. 1992) and that, although the quantity of material in the Martian atmosphere varies with the seasons, it never drops entirely to zero. As on Earth, mineral dust affects the radiation balance of the Martian atmosphere, its thermal structure and atmospheric circulation (Leovy 2001). Dust in the Martian atmosphere, for example, reduces diurnal temperature variations near the ground (Read and Lewis 2004).

Wind is the most active geomorphological agent currently operating on Mars and it seems likely that dust has had long-term effects on the planet's surface. The omnipresence of dust in the atmosphere is also a potential hazard for any instrumentation delivered to the Martian surface for *in situ* analyses. Suspended particles readily adhere to all types of surfaces causing optical obscuration and potential damage to both mechanical and electrical systems (Landis and Jenkins 2000). This problem is made acute by the chemical activity of Martian dust, which is thought to be highly oxidizing (Plumb et al. 1989), although its mineralogy is not adequately known.

5 The Regional Picture

5.1 Introduction

In this chapter we discuss the regional geography of dust storms in the world's drylands, using available climatological data and information from the Total Ozone Mapping Spectrometer (TOMS) and other types of remote sensing. We discuss such issues as annual frequency, seasonality and transport trajectories.

5.2 North America

Dust-raising is a common feature of the dryland climate characteristic of large parts of the United States (Fig. 5.1), but meteorological observations at ground stations suggest that the greatest frequency of dust events occurs in the panhandles of Texas and Oklahoma, Nebraska, western Kansas, eastern Colorado, the Red River Valley of North Dakota (Godon and Todhunter 1998) and northern Montana (Fig. 5.2). These areas combine erodible materials with a moderately dry climate and high values for wind energy (Gillette and Hanson 1989). The spring months are the time of maximum dust activity over much of the area (Orgill and Sehmel 1976; Stout 2001) but a summer peak occurs in parts of Arizona (Brazel 1991) due to convective or thunderstorm activity. In Late Pleistocene times dust storm activity may have been even greater, leading to loess deposition (Muhs et al. 1999).

Large amounts of dust are also blown off the bed of the former Owens Lake following its anthropogenically caused desiccation (Reheis 1997) and from shrunken Mono Lake (Gill and Cahill 1992). However, land management techniques are probably important in determining the variability of dust storm occurrence. Lee et al. (1993) and Todhunter and Cihacek (1999) document a decline in dust storm occurrence in Texas and North Dakota, which they attribute to the adoption of improved land-use practices. A discussion of the spatial and temporal variability of dust storms in the Mojave and Colorado deserts is provided by Bach et al. (1996), who identify the Coachella Valley as being the dustiest region. A detailed study of dust deposition in Nevada and California is provided by Reheis and Kihl (1995).

Fig. 5.1. A dust storm at Zabriskie Point, Death Valley (Calif., USA; from ASG)

From time to time, dust events are recorded further to the east in the United States, as has happened during drought years like 1981 in Illinois (Changnon 1983).

The TOMS data show only one area with maximum AI values greater than 5.0 – parts of the Great Basin. This is an area of fault-bounded blocks and troughs which contains over 150 basins separated from each other by north–south trending mountain ranges. Most of the basins were occupied by Pleistocene lakes that covered an area at least 11 times greater than the area they cover today (Grayson 1993, p. 86). One of these was Bonneville, which was roughly the size of present-day Lake Michigan. Another was Lahontan, which covered an area roughly as great as present-day Lake Erie. Their desiccation, the presence of extensive areas of salty lake floor (Blank et al. 1999) and the existence of large expanses of alluvial fans running into the many basins may account for the importance of this area as a dust source.

Much of the dust in the High Plains may occur at low levels and so may be inadequately detected by TOMS. Three synoptic patterns are associated with dust events in the southern High Plains (Wigner and Peters 1987). One of these is convective modification of the boundary layer. This accounts for 42% of dust events at Lubbock and causes strong winds at low levels, particularly in late morning (Lee et al. 1994). Another 19% of all events are caused by thunderstorm outflows, which again may have a limited vertical extent. The passage of cold fronts usually limits the vertical spread of dust.

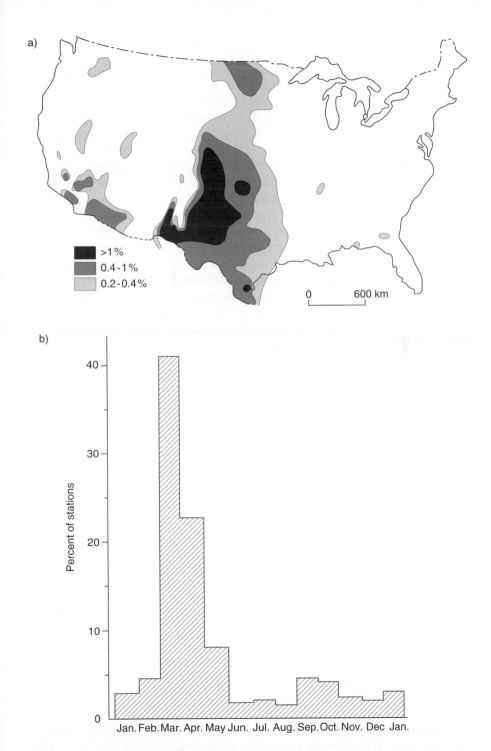

Fig. 5.2. Dust storms in the United States. a) Annual frequency of dust hours with visibility less than 11 km. b) The percentage of observation stations that have a maximum frequency of airborne dust during a particular month. Modified after Gillette and Hanson (1989, Fig. 8) and Orgill and Sehmel (1976, Fig. 3)

In addition, while the map of dust storm occurrence in the United States is based on the work of Orgill and Sehmel (1976), the TOMS data relate to an entirely different and more recent period. Over that time, changes in land use have caused a decrease in dust storm activity in some areas, including the Lubbock region of Texas (Ervin and Lee 1994) and North Dakota (Todhunter and Cihacek 1999).

Possibly the most famous case of soil erosion by deflation was the Dust Bowl of the 1930s in the United States. In part this was caused by a series of hot, dry years associated with anomalous sea surface temperatures (Schubert et al. 2004) which depleted the vegetation cover and made the soils dry enough to be susceptible to wind erosion. The effects of this drought were gravely exacerbated by years of intensive grazing and unsatisfactory farming techniques. However, perhaps the prime cause of the event was the rapid expansion of wheat cultivation in the Great Plains (see Chapter 7). The dust settled at great distances from source, including Canada, New Hampshire (Page and Chapman 1934), Illinois (Van Heuklon 1977), Philadelphia (Watson 1934), New York (Anon 1935), Wisconsin (Winchell and Miller 1918) and Louisiana (Russell and Russell 1934).

Dust storms are still a serious problem in various parts of the United States: the Dust Bowl was not solely a feature of the 1930s. Thus, for example, in the San Joaquin Valley area of California in 1977, a dust storm caused extensive damage and erosion over an area of about 2000 km^2. More than 25×10^6 t of soil were stripped from grazing land within a 24-h period. While the combination of drought and a very high wind (as much as 300 km h^{-1}) provided the predisposing natural conditions for the stripping to occur, intensive grazing and the general lack of windbreaks in the agricultural land played a more significant role. In addition, broad areas of land had recently been stripped of vegetation, levelled or ploughed up prior to planting. Elsewhere in California dust yield has been considerably increased by mining operations in dry lake beds (Wilshire 1980).

Dust storms also occur in the Canadian Prairies (Wheaton and Chakrabarti 1990), most notably in southern Saskatchewan, where they occur on average on over five days per year (Fig. 5.3a). The peak season for dust storms is the spring, when wind speeds are high, vegetative cover is sparse and precipitation amounts are lower than in the summer (Fig. 5.3b). They can cause considerable loss of top soil (Wheaton 1992) and were particularly virulent in the 1930s (Wang 2001).

Large areas of Mexico have a dryland climate but the study of dust storms has been concentrated in only a few locales, notably Mexico City (Jauregui 1960, 1989; see also Chapter 7). Remote sensing imagery suggests dust-raising may be widespread (Fig. 5.4); and the country's northern deserts have been identified as important contributors to eastern Pacific sediments (Bonatti and Arrenhius 1965). Inputs of dust from Mexican deserts have also been identified in soils locally (Slate et al. 1991) and in the United States south-west (Reynolds et al. 2003).

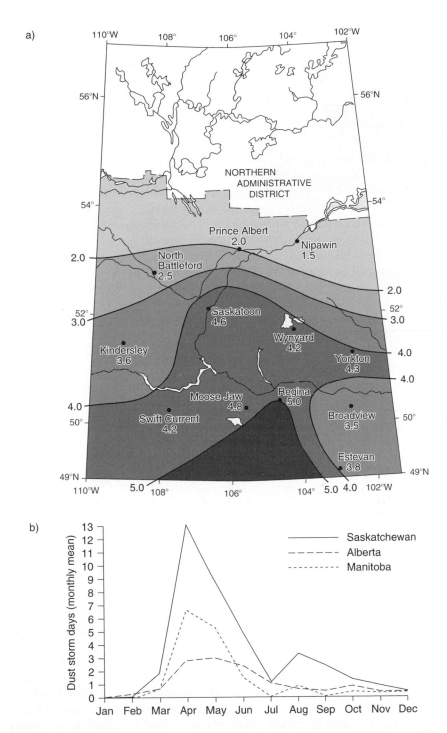

Fig. 5.3. a) Annual dust storm frequency (days), Saskatchewan (1977–1983). b) Seasonal distribution of the mean frequency of days with dust storms for each prairie province (1977–1985). Modified after: (a) Wheaton and Chakrabarti (1987, Fig. 4), (b) Wheaton and Chakrabarti (1990, Fig. 6)

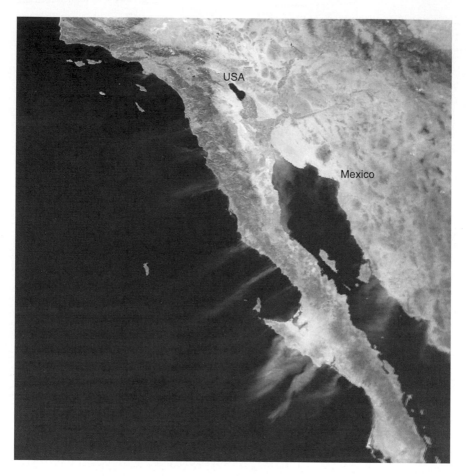

Fig. 5.4. Multiple dust plumes from Baja California, 10 February 2002 (Seawifs)

5.3 South America

Information on the occurrence of dust storms in South America is sparse. However, Johnson (1976) suggests that dust storms are frequent in the Altiplano of Peru, Bolivia and Chile. In Iquique (Chile), most dust activity occurs in the winter months (May to September). Middleton (1986c) noted their importance on the Puna de Atacama where salt basins – *salars* – appear to be important sources. The presence of extensive areas of closed depressions and of wind-fluted topography, combined with the probable importance of salt weathering in the preparation of fine material for deflation (Goudie and Wells 1995), suggest that the dry areas of the Altiplano should indeed be major source areas for dust storms. TOMS identifies one area in South America where aerosol values are relatively high. This is the Salar de

Uyuni, a closed basin in the Altiplano of Bolivia which is located in an area with 178 mm of annual rainfall. This salar, the largest within the Andes, is possibly the world's largest salt flat, though in the Late Pleistocene it was the site of a huge lake, pluvial Lake Tauca (Lavenu et al. 1984). The pluvial lake was more than 600 km long; and it is likely that the fine sediments from its desiccated floor are one of the reasons for the existence of high aerosol values in this region. It is of the same order of size as some of the other major basins that are major dust sources (e.g. Bodélé/Chad, Eyre, Taklamakan and Mkgadikgadi).

In addition, as Middleton (1986c) has shown, there is a tract to the west of Buenos Aires in Argentina where dust storm activity is substantial, with extensive areas experiencing more than eight dust storms per year (Fig. 5.5). In Mendoza, where there is a winter (Mediterranean) rainfall regime, the dustiest months run from April to August. Rates of dust accretion in the Pampas are appreciable at the present day (Ramsperger et al. 1998). In addition, this arid zone has the most extensive spreads of loess in the southern hemisphere (Teruggi 1957; Kröhling 1999; Sayago 2001). Large numbers of closed depressions attest to the power of deflation. Dust trapped in the West Antarctic glaciers and on surfaces in the South Shetland Islands may have a Patagonian origin (Iriondo 2000; Gaiero et al. 2004; Lee et al. 2004).

5.4   Southern Africa

Southern Africa is not a major area of dust production from a global perspective, even though it has large areas of arid terrain both in the coastal Namib and in the interior (Kalahari and Karoo). There are extensive areas of pans (Goudie and Wells 1995), many of which are, at least in part, the result of deflation; and there are many windstreaks and yardangs known from the Namib. Examination of satellite images has shown the presence of dust plumes driven by *Berg* winds blowing westwards off the Namib (Fig. 5.6) and the Kalahari towards the South Atlantic (Eckardt et al. 2002). In addition, sedimentological studies have shown the presence of loess and loess-like deposits in parts of Namibia (Eitel et al. 2001).

TOMS analyses indicate that there are two relatively small, but clearly developed dust source areas in southern Africa (Fig. 5.7). The most intense of these is centred over the Etosha Pan in northern Namibia (Bryant 2003). The other centre is over the Mgkadikgadi Depression in northern Botswana.

The Etosha Pan, which covers an area of about 6000 km^2, comprises a salt lake that occupies the sump of a much larger fault-controlled basin. The salt lake often floods in the summer, but is for the most part dry enough in the winter for deflation to occur, as is made evident by the presence of extensive lunette dunes on its lee (western) side (Buch and Zoller 1992). It is fed by an extensive system of ephemeral flood channels – *oshanas* – that have laid

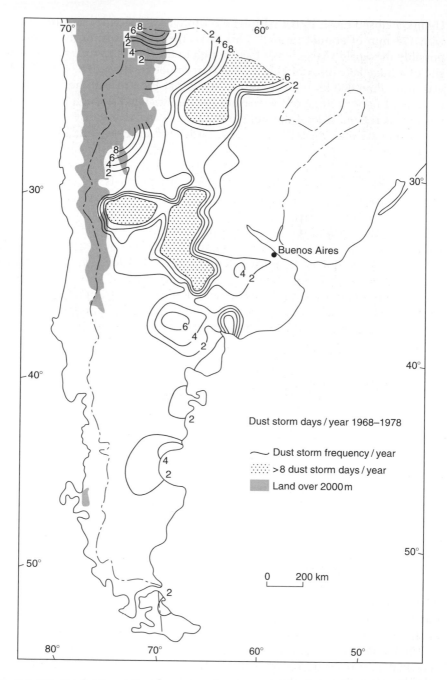

Fig. 5.5. The distribution of dust storms in Argentina. Note that no visibility limit is used. From Middleton (1986c, Fig. 11.15)

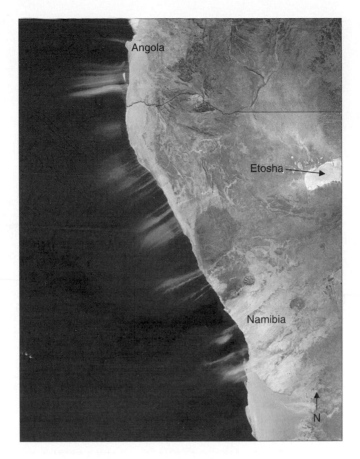

Fig. 5.6. Plumes of dust blowing off the Namib Desert of Namibia and southern Angola, 9 June 2004 (MODIS)

down large tracts of susceptible fine-grained alluvial and lacustrine sediments. In the past it is possible that it has also received large inputs of material from the highlands of Angola via the Cunene River (Wellington 1938). Flood events have a marked effect on immediate and subsequent dust emissions (Bryant 2003).

The Mkgadikdadi depression of northern Botswana is another major structural feature, the floor of which is now occupied by a series of saline sumps. In dry years these present surfaces from which deflation can and does occur. The pans are, however, but shrivelled remnants of a former pluvial lake, Lake Palaeo-Mkgadikgadi, which at its greatest extent covered 120 000 km^2 (Thomas and Shaw 1991). It was second in Africa only to Lake Chad at its Pleistocene maximum. It was fed with water and sediment from the Okavango and, perhaps, Zambezi systems and by more locally derived rivers (*mekgacha*) flowing from the south.

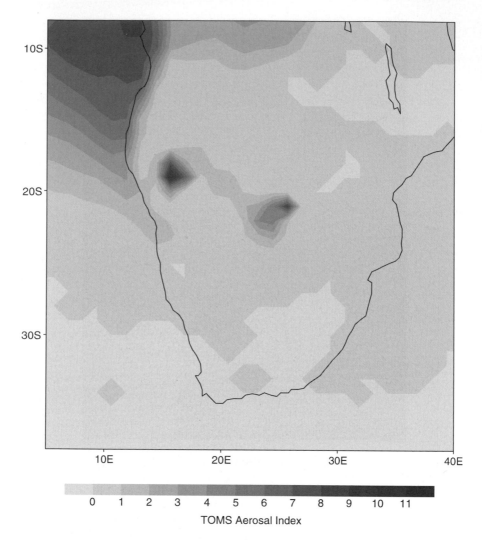

Fig. 5.7. Annual average TOMS aerosol index (AI) values for southern Africa. Modified after Washington et al. (2003, Fig. 14)

Dust events in the southern African source regions are invariably associated with enhancement of the low-level easterly circulation over the interior (Eckardt et al. 2002). Transient eddies, in the form of west-to-east migrating anticyclones travelling to the west of a Rossby wave-trough axis, are confined to the oceanic areas immediately to the south of the subcontinent as a result of the unbroken escarpment (De Wet 1979; Tyson and Preston-Whyte 2000). The migration of mass in these systems leads to an enhanced east–west gradient and the corresponding anomalous easterlies, which, over the western half of the subcontinent, are associated with dust storms and plumes over the subtropical southeast Atlantic. One particularly intense storm that transported large amounts of material across the interior of South Africa, probably originating in the Mkgadikdadi salt pans, occurred in August 2003 (Resane et al. 2004).

Figure 5.8 shows an overlay of TOMS values, potential sand flux (q) and elevation derived from a digital elevation model at 0.5° resolution. Data for TOMS AI values and for potential sand flux relate to July–September, which corresponds with the season of largest AI values in Etosha and Mkgadikgadi. Unlike the cases of the AI maxima in the Sahara and China, there is no clear association between a maximum of potential sand flux and AI values. Neither of the two pans is located in a region where topographic channelling of the wind would accelerate it sufficiently to produce a large dust source.

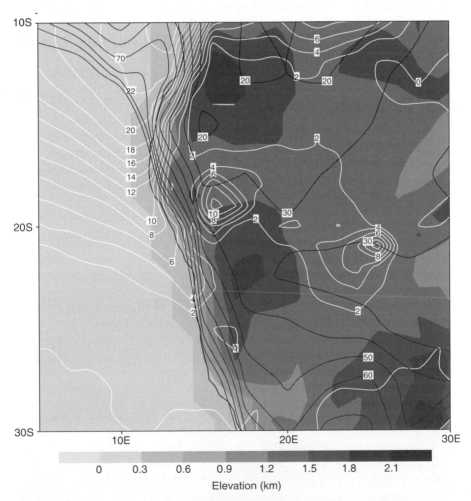

Fig. 5.8. TOMS aerosol index (AI) values (white contours, contour interval 2), scaled potential sand flux (black contours, contour interval 10) and elevation in km (shaded) for southern Africa, long-term means, July–September. Modified after Washington et al. (2003, Fig. 15)

The region is remarkably flat. Instead, it is likely that the southern Africa dust sources are supply-limited, with suitable material available only from the two pans (Washington et al. 2003). Much of the rest of northern Botswana and northern Namibia, which is relatively moist, with mean annual precipitation levels of around 400–800 mm, is covered by savanna woodland and grassland, and so is not readily susceptible to deflation.

5.5   The Sahara

The Sahara and its margins are the world's major source of aeolian mineral aerosol (Morales 1979; Brooks 1999; Kaufman et al. 2005; Fig. 5.9). This amounts to almost half of all aeolian desert material supplied to the world's oceans. There are many different dust-bearing winds in the region (Table 5.1); and the power of these winds as deflational agents is particularly great north of 15° (Clark et al. 2004). On the south side of the Sahara winds exceed the threshold velocity for sediment particle movement during two distinct seasons. During the dry season from October to April the area is subject to north easterly trade winds, locally called *Harmattan*. In the early rainy season the atmosphere is very unstable and strong convectional activity occurs. Fully developed thunderstorms associated with large cumulonimbus clouds produce strong vertical downdrafts that cause a vigorous forward outflow of cold, dust-raising air.

One of the most important needs in furthering our understanding of the Saharan production of dust is to identify the major source areas (Stuut

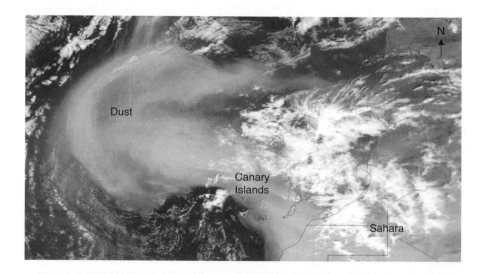

Fig. 5.9. A Seawifs image of a Saharan dust storm on 26 February 2000

Table 5.1. Local names for Saharan dust-bearing winds (after Middleton 1986c)

Name (derivation when known)	Area affected	Season	Direction from	Meteorological conditions
Irifi	Western Sahara		SE	Frontal
Ghibli (free translation: 'wind from south Mecca')	Tripolitania			Pre-frontal
Guebli (south wind)	Tunisia and Algeria (northern parts)	All year, but most prevalent May–October	S	Pre-frontal with katabatic effects from interior uplands to coastal plains
Sahel	Morocco		SW	Frontal
Harmattan (Fantee: 'aharaman' to blow and 'ta' grease locally, used to cover skin)	Bilma/Faya Largeau area plus much of 20° N	October–April	ENE	Pressure surge after cold air outbreaks from mid-latitudes
Brume sèche (French: 'dry haze')	West Africa	October–April		Harmattan haze in light winds
Haboob (Arabic: 'to blow')	Sudan (but has become almost generic in its use)	May–July		Single-cell thunderstorm downdraft
Khamsin (Arabic: 'fifty'[a])	Egypt	Spring		Pre-frontal
Chili	Tunisia and Algeria	Spring	SW	Pre-frontal
Shekheli	Algeria	Spring		
Chergui	Moroccan Sahara	Summer	NE	
Dschani	Southern Sahara			
Kharif	Somalia	June–September	SW	
Gobar	Ethiopia			
Sirocco	Southern Europe	Spring	S	Frontal
Leveche	Spanish Mediterranean coast: Malaga–Alicante	Spring	SE–SW	Frontal
Leste	Maderia			Frontal
Levanto	Canary Islands			

[a] Variously taken to refer to the average duration of the wind (50 h), its annual frequency (50 times) and its season of maximum onset (the 50 days either side of the spring solstice)

et al. 2005). Hermann et al. (1999) attempted to survey previous studies of Saharan dust sources but remarked (p.142): "Astonishingly, the results with regard to source areas are totally different. No overlapping can be detected which could serve as a confirmation of results". One reason for this situation is the range of source identification methods that have been used by different workers: remote sensing, analysis of surface dust observations, back trajectory analysis of isobar data and the use of mineral tracers. This lack of agreement over source areas is evident from comparison of various attempts at their delimitation shown in Fig. 5.10.

However, in recent years, some progress has been made in identifying source areas by measurements of infra-red radiances such as those acquired by METEOSAT. These can be used to produce the Infra-red Difference Dust Index (IDDI; Brooks and Legrand 2000). This method has highlighted the Bodélé Depression between Tibesti and Lake Chad (centred on 16° N and 18° E) as an important source region throughout the year, together with a large swathe of country covering portions of Mauritania, Mali and southern Algeria. It also suggests that the Horn of Africa (see also Léon and Legrand 2003) and the Nubian Desert in southern Egypt and northern Sudan are important sources. The importance of the Bodélé region was also shown by Kalu (1979) and Hermann et al. (1999), but the status of the other regions is less clear.

5.5.1   Saharan Sources

TOMS data (Fig. 5.11) confirm that the Bodélé is the most intense source region, not only in the Sahara, but also in the world (Giles 2005), with AI values that exceed 30. It also demonstrates the presence of a large but less intense area (AI values over 24) in the West Sahara. This extends through to the Atlantic coast of Mauritania. Relatively high AI values are also observed in the interior of Libya, where numerous dust plumes have been observed with SeaWifs (Koren et al. 2003).

Because of the high correlation between TOMS AI values and dust optical thickness (DOT) as determined by sun photometer readings, it is possible to construct maps of DOT (and thus total atmospheric dust load) over the Sahara (Moulin and Chiapello 2003; Fig. 5.12). This highlights the importance of the zone between 15° N and 22° N and confirms the high level of dustiness developed over the western Sahara.

The importance of Bodélé as a dust source is related to various factors. First, the region is very dry (Faya Largeau receiving an average annual rainfall of just 17 mm), but is fed with silty alluvium by streams draining from the Tibesti Massif. There may also be susceptible silty materials that were laid down in an expanded Lake Chad during early Holocene and Pleistocene pluvials, together with extensive spreads of ancient diatomites, many of which are furrowed by yardangs. In addition, Mainguet and Chemin (1990) have argued that deflational activity downwind from Tibesti may be substantial

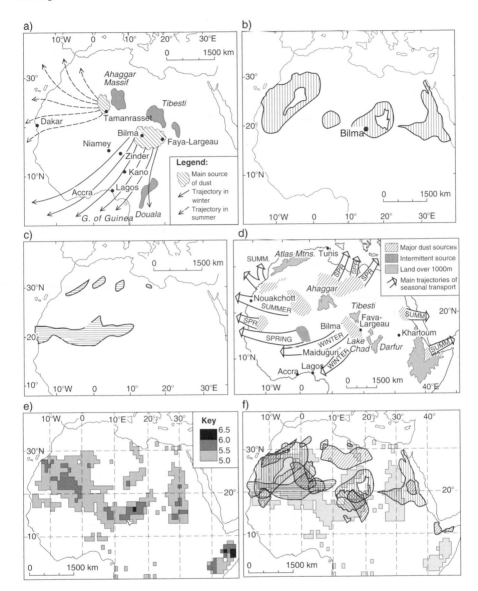

Fig. 5.10. Previous attempts to delimit Saharan dust source areas. a) After Kalu (1979), b) after D'Almeida (1986), c) after Dubief (1953) d) after Middleton (1986), e) after Brooks and Legrand (2000), f) composite of a–e. Modified after Middleton and Goudie (2001, Fig. 1)

and help to explain the excavation of Lake Chad itself. There is distinct topographic funnelling of high velocity winds. Moreover, Washington and Todd (2005) have pointed to the importance of the Bodélé low-level jet in creating dust emissions from the area. This is a feature which uniquely overlies the greater Bodélé region rather than areas surrounding it.

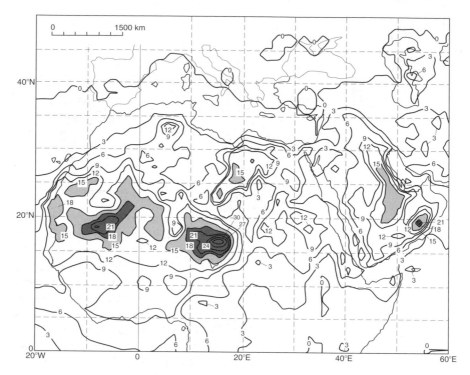

Fig. 5.11. Mean aerosol index values for Northern Africa and the Middle East from TOMS data (1980–1993, 1997–2000). Modified after Middleton and Goudie (2001, Fig. 4)

The reasons for the importance of the West Saharan dust source in Mali, Mauritania and Algeria are less well understood. However, it is an area of low relief bounded on the north and east by uplands. While such upland areas are not themselves major dust source regions, ephemeral wadis draining from them have transported silt-rich alluvium into the area. Likewise, in the past, the southern part of the region may have received alluvial inputs from the Niger River prior to its capture by southeast-trending drainage near Tosaye (Urvoy 1942). It also contains an enormous closed depression some 900 km long and various ergs that could provide a dust source through winnowing. The depression contains many ancient lakebeds that show signs of intense deflation in the Holocene (Petit-Maire 1991). Dubief (1953) maps it as an area of high aeolian activity; and it is also rather dry, with annual precipitation levels of 5–100 mm.

Interestingly, both of these two main source areas are little affected by anthropogenic activities. They have very few settlements and are too dry to support settled agriculture, each having an extremely arid climate. Although several studies have shown the importance of areas in Sahelian latitudes as source areas that have increased their dust output during periods character-ized both by prolonged drought and intensified land use, in places leading to

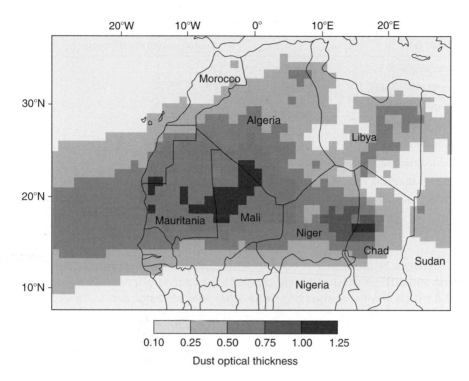

Fig. 5.12. The mean summer DOT over Africa and the North Atlantic (1979–2000) determined by TOMS. Modified after Moulin and Chiapello (2003, Fig. 2a)

desertification (Goudie and Middleton 1992; N'Tchayi et al. 1997), the Sahara's two major dust sources are primarily driven by natural climatic and geomorphological factors.

However, the relative lack of human activity in these two major Saharan dust source regions also means a relative dearth of ground-based information on the wind erosion system in these areas. Data derived from remote-sensing platforms has helped to fill these information gaps in recent times. Koren and Kaufman (2004) estimated that a minimum wind speed of 10 m s^{-1} is needed to initiate a dust storm in the Bodélé depression following their examination of some 15 storms in 2003 with Moderate resolution imaging spectroradiometer (MODIS) data from the Terra and Aqua satellites. Since the Aqua passes over the depression around 3 h after Terra, the authors were also able to monitor the movement of dust plumes from the Bodélé. Their analysis indicates that the dust clouds are blown along at around 13 m s^{-1}, which is about double the speed calculated from previous ground-based measurements.

Various attempts have been made to estimate and model the source strength of the Sahara, using data on mineral loadings in the atmosphere, surface material characteristics (Callot et al. 2000) and transport models (Table 5.2). The estimates show a wide range of values that may reflect

Table 5.2. Estimates of the source strength of the Sahara

Author (s)	Annual quantity ($\times 10^6$ t year^{-1})
Jaenicke (1979)	260
Schütz et al. (1981)	260
Prospero (1996a, b)	170
Swap et al. (1996)	130–460
D'Almeida (1986)	630–710
Marticorena and Bergametti (1996)	586–665
Callot et al. (2000)	760
Ginoux et al. (2004)	1400
Miller et al. (2004)	517

differences in modelling procedures, differences in the time-scales considered and differences in the areal extent of the source. There are few data available which allow a comparison with other major source areas. An exception to this is provided by the work of Zhang et al. (1997) on the Taklamakan Desert. For this region, they estimate an annual dust production of 800×10^6 t. On this basis, they propose that this may be around half of the global production of dust. Other data are presented in Table 5.3.

An alternative method that can be used to compare relative source strengths is the TOMS data. By looking at the AI intensity and its areal extent, it is possible to gain an indication of the predominance of the Sahara in comparison with other desert areas. As Table 4.2 shows, three of the world's four most important dust sources occur within the Sahara, and Fig. 5.13 shows the area and intensity of the Saharan AI compared with those for Arabia, northwest China and the Thar. It confirms the importance of the Sahara on the global scale.

The seasonal patterns show great variability. Dust activity appears to be very limited over the entire area in October, November, December (OND; Fig. 5.14a). There is minimal activity in the eastern and northern parts and relatively little dust presence over the North Atlantic. The only area that has a substantial number of days (27%) with AI values over 19 is the Bodélé depression, which is active throughout the year. Aerosol presence is greater in January, February, March (JFM; Fig. 5.14b), especially in the southernmost part of the area, but it is possible this may partly be the result of smoke from biomass burning during the dry season. However, once again Bodélé emerges as that with the highest frequency of dust. April, May, June (AMJ; Fig. 5.14c) is the period with the greatest level of dust occurrences, with much of the region being active. Three squares have AI values of over 19 on more than 80% of days; and there is a large area, centred on Mauritania and Mali, where this is the case for over 60% of days. There is also a moderate level of activity over the Middle East and on the southern borders of the Mediterranean.

Table 5.3. Estimates of global and regional dust emissions

Region	Tg year^{-1}	%
Global emissions (derived from data in Miller et al. 2004, Fig. 1)		
Sahara/Sahel	–	50.7
Central Asia	–	16.0
Australia	–	14.5
North America	–	5.2
East Asia	–	4.9
Arabia	–	4.2
Others	–	4.5
Dust emissions (derived from data in Ginoux et al. 2001)		
North Africa	1430	69.0
South Africa	322	1.1
North America	9	0.4
South America	55	2.7
Asia	496	23.9
Australia	61	2.9
Total emissions	2073	100
Global Emissions in 1998 (derived from data in Luo et al. 2003)		
North Africa	1114	67.4
Arabian Peninsula	119	7.2
Asia	54	3.3
Australia	132	8.0
Miscellaneous	235	14.2
Total emissions	1654	100

In July, August, September (JAS; Fig. 5.14d), the Mauritania–Mali area continues to dominate, but the southern part of the region (probably as a reflection of the main rainy season in tropical West Africa) is relatively inactive. It is a season when dust occurrence over the western Mediterranean is at its highest and when there are major dust deposition events in Corsica, Sardinia and their neighbours.

The seasonal pattern of dust activity in the Sahara can also be assessed through analysis of meteorological data. In Table 5.4 we present data on the percentage number of days with blowing dust/sand for each month for a latitudinal sequence of stations from Abidjan (Ivory Coast) in the south to Sousse (Tunisia) in the north. The two months in the year with the highest percentages of dust days are highlighted. In the south [essentially between Abidjan and Dakar (Senegal)], the highest percentages occur between November and March. This is the time of the *Harmattan*. By contrast, as we move into the central and northern Sahara, April to August is the time with the highest percentages. In other words, there is an annual migration of the dust centre of gravity over the course of the year, a finding that is confirmed by MODIS observations (Kaufman et al. 2005).

One area in northern Africa that repeatedly generates dust, particularly in the dry summer months (June and July), when *haboobs* are common (Tothill

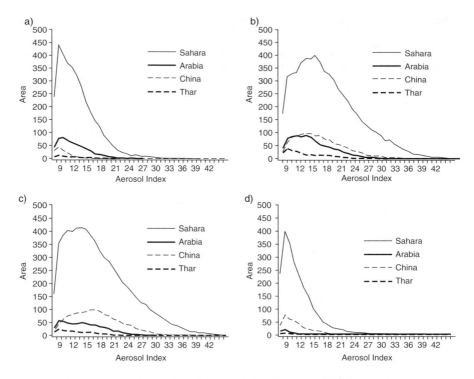

Fig. 5.13. The extent and intensity of the AI derived by TOMS for four major desert areas: Sahara, Arabia, China and Thar. The figure shows the areas (in km^2 × 10^3) covered by different intensities of the AI. Modified after Goudie and Middleton (2001, Fig. 2)

1948, p. 709), is the Tokar Delta area of Sudan (18.5° N, 37.7° E; Fig. 5.15). This 2150 km^2 delta, fed by the silt-laden Baraka River, is located on the Red Sea coast approximately 170 km south of Port Sudan and is in an arid area (mean annual precipitation of around 74 mm). The delta consists of alluvial silts across which winds are funnelled through a major gap (ca. 110 km wide) in the Red Sea Hills. MODIS imagery pinpoints this hot spot for dust generation repeatedly (Fig. 5.16). Dust storms are also common over parts of Egypt; and Table 5.5, based on ground observations, provides information on the distribution and frequency of dust storms in that country (Banoub 1970). Dust storm activity appears to be greatest between December and May.

5.6　Trajectories of Saharan Dust Transport

Saharan dust is regularly transported from its source areas along four main transport paths: (a) southwards to the Gulf of Guinea and to countries such as Ivory Coast and Ghana (Breuning-Madsen and Awadzi 2005), (b)

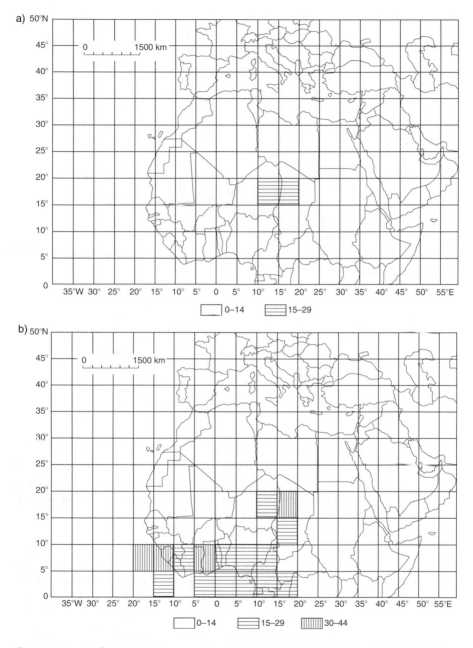

Fig. 5.14. Seasonal 1999 TOMS maps. Percentage of days with AI values >1.9: a) October, November, December (OND), b) January, February, March (JFM),

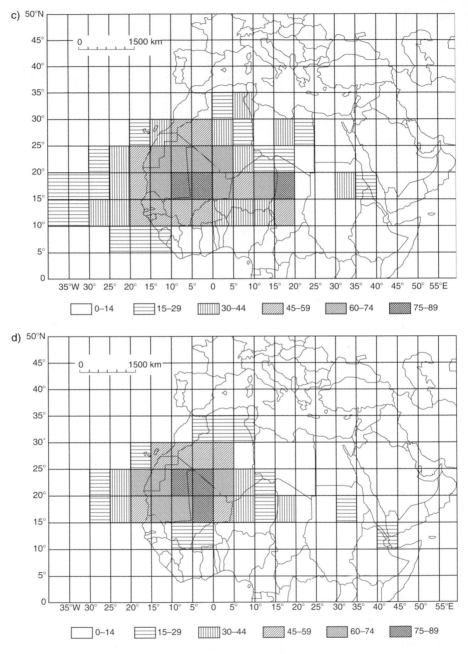

Fig. 5.14. (*continued*) c) April, May, June (AMJ), d) July, August, September (JAS). Modified after Middleton and Goudie (2001, Fig. 3)

Table 5.4. Seasonality of blowing sand/dust in North Africa. Source: Weatherbase.com

Location (latitude)	J	F	M	A	M	J	J	A	S	O	N	D
Abidjan (5° 15′ N)	45.5	9.1	9.1	–	–	–	9.1	–	–	–	–	27.3
Zaria (11° 08′ N)	15.9	12.6	12.6	7.3	2.0	2.0	1.3	3.3	3.3	9.9	15.2	14.6
Maiduguri (11° 51′ N)	16.1	13.3	15.0	11.1	3.9	1.1	1.1	0.6	1.1	7.2	13.9	15.6
Sokoto (13° 01′ N)	13.2	12.0	15.0	11.4	6.0	2.4	1.8	1.2	1.8	7.8	14.4	13.2
Niamey (13° 29′ N)	13.7	11.8	14.3	11.8	7.5	4.3	1.9	1.2	2.5	8.1	10.6	12.4
Zinder (13° 47′ N)	11.8	10.8	11.8	10.3	8.7	7.2	3.1	1.0	4.1	9.7	10.8	10.8
Mopti (14° 31′ N)	13.6	12.8	14.4	12.0	8.0	8.0	4.0	0.8	2.4	6.4	6.4	11.2
Dakar (14° 44′ N)	15.7	13.0	14.8	13.0	9.3	3.7	1.6	0.9	2.8	6.5	7.4	11.1
Timbuktu (16° 43′ N)	9.0	7.8	10.1	10.1	9.6	11.7	9.6	6.4	6.4	6.4	4.8	7.8
Agadez (16° 58′ N)	8.1	7.7	9.9	9.0	10.8	10.8	10.7	6.3	5.9	6.8	7.2	7.2
Nouakchott (18° 06′ N)	9.2	8.4	9.9	9.9	10.6	9.5	7.7	6.2	7.0	7.3	6.6	7.7
Bilma (18° 41′ N)	8.3	8.3	10.6	9.3	9.3	9.3	10.2	8.8	6.9	6.0	6.0	6.9
Atar (20° 31′ N)	8.1	6.7	9.1	7.7	7.7	9.6	12.4	12.0	9.6	5.7	5.3	6.2
Nouadhibou (20° 56′ N)	7.7	7.3	9.8	9.8	10.6	10.6	10.2	8.9	8.1	6.1	5.3	5.7
Ad Dakhla (23° 42′ N)	8.4	6.7	7.6	4.2	6.7	8.4	13.4	14.3	8.4	6.7	6.7	8.4
Sousse (35° 46′ N)	9.7	7.5	6.0	9.0	9.0	9.7	17.2	12.7	6.7	3.7	7.5	1.5

a)

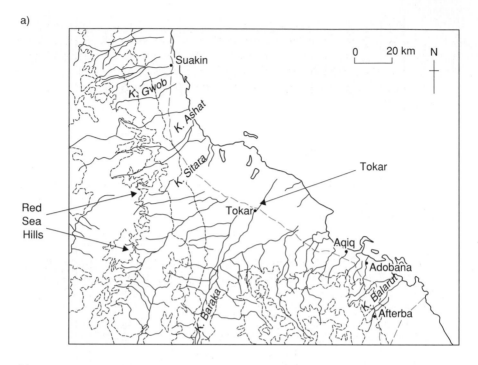

b)

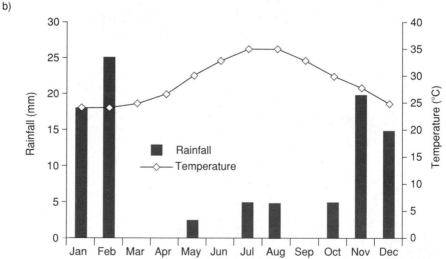

Fig. 5.15. a) The location of the Tokar delta, Sudan. b) Temperature and rainfall data for Tokar

westward over the North Atlantic Ocean (NAO; Carlson and Prospero 1972:
Moulin et al. 1997) to the islands of the eastern Atlantic such as the Canaries
(Alastuey et al. 2005), North America (Perry et al. 1997) and South America
(Swap et al. 1992), (c) northward across the Mediterranean (Löye-Pilot et al.

1986) to southern Europe, sometimes as far north as Scandinavia and the Baltic (Franzen et al. 1994; Papayannis et al. 2005; Barkan et al. 2005) and (d) along easterly trajectories across the eastern Mediterranean (Herut and Krom 1996; Kubilay et al. 2000, 2005) to the Middle East (Ganor et al. 1991) and possibly as far as the Himalayas (Carrico et al. 2003) and East Asia as far as Japan (Tanaka et al. 2005).

5.6.1   North Atlantic Trajectories

The westward flow of material over the NAO is the most voluminous, accounting for 30–50% of output (Schütz et al. 1981; D'Almeida 1986). Numerous papers have documented the transport and deposition of Saharan

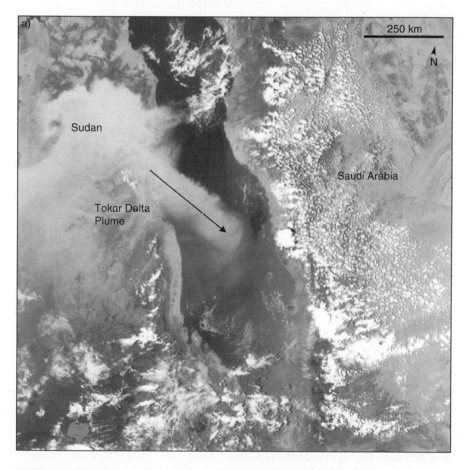

Fig. 5.16 Two MODIS images of the Tokar delta, bordering the Red Sea in Sudan, showing plumes of dust blowing westwards towards Saudi Arabia. a) 11 July 2002, b) 1 September 2004

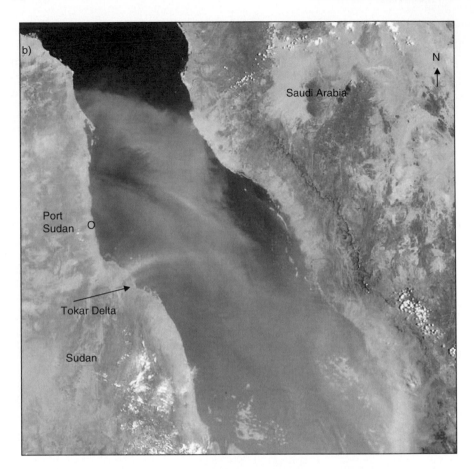

Fig. 5.16. (*continued*)

dust to distant regions of the NAO and to the Americas (for reviews, see Duce 1995; Prospero 1996a). Kaufman et al. (2005), using MODIS, calculated that around 240 Tg of dust are transported annually into and across the Atlantic Ocean, with 50 Tg of this fertilizing the Amazon Basin.

In his voyage on the *Beagle*, Charles Darwin reported that the atmosphere was generally hazy over the Cape Verde Islands and stated (1893, p. 18):

"I have found no less than fifteen different accounts of dust having fallen on vessels when far out in the Atlantic. From the direction of the wind whenever it has fallen, and from its having always fallen during those months when the harmattan is known to raise clouds of dust high into the atmosphere, we may feel sure that it comes from Africa".

Large dust outbreaks during the summer appear to be associated with strong convective disturbances that develop over West Africa at about 15–20° N

Table 5.5. Dust storm frequencies in Egypt (1964–1968). Processed from data provided by Meteorological Dept, Cairo, Arab Republic of Egypt

Location	Annual frequency	
	>1000 m visibility	<1000 m visibility
Sallum	40	10.0
Mersa Matruh	38	9.6
Alexandria	13	3.4
Port Said	12	1.6
Tanta	0.5	0.2
Cairo	46	6.5
Fayum	5	0
Minya	17	0
Assyout	33	1.6
Luxor	32	2.2
Aswan	68	6.6
Siwa	25	1.4
Baharija	16	2.4
Farafra	32	2.8
Daklha	47	0.6
Kharga	49	2.0
Hurghada	23	2.0
Qusier	30	1.6
Mean value	29.2	3.0

and move westward, carrying material entrained in Saharan and Sahelian latitudes. Resultant dust plumes over the NAO are usually associated with easterly waves that emerge from the African coast every 3–4 days. Their complex structure produces intricate distribution patterns, including northward branches that can transport material to Western Europe. Remote sensing images over the NAO have also demonstrated the importance of the development of the Azores–Bermuda high-pressure system in summer in drawing dust-laden air from the tropical North Atlantic into the subtropical region (Jickells et al. 1998).

Saharan dust outbreaks over the NAO commonly start over North Africa in a deep, well mixed, dry adiabatic layer of air that is undercut by cool, moist low-level air as it advances westward and emerges from the African coast to become a discrete Saharan air layer (SAL). The SAL, which is usually characterized by a temperature inversion at its base, is often associated with a mid-level easterly jet. Outbreaks of the dust-laden SAL commonly persist for several days but exceptionally can last for tens of days (Carlson 1979).

The longer-lived plumes transport material the furthest. Transport to the Caribbean, where an estimated 20×10^6 t of Saharan dust is deposited annually (Schlatter 1995), typically takes 5–7 days (Prospero and Carlson 1981). The duration of individual Saharan dust events, monitored at Trinidad in the West Indies, can vary between 3 days and 5 days; and sometimes back-to-back episodes can last as long as 20 days (Rajkumar and Chang 2000).

The latitudinal pathways of transatlantic transport vary seasonally. Hence, the maximum Saharan aerosol concentrations monitored at Barbados and Miami are in July and August (Prospero and Carlson 1981; Prospero et al. 1987; Prospero 1999), while the highest concentrations monitored at Cayenne (Prospero et al. 1981) are in March (Fig. 5.17). Sal Island lies in a zone that is affected by both of these seasonal pathways, displaying a bimodal peak (March and August/September) in atmospheric turbidity (Schütz 1979). TOMS analysis shows this clear seasonal pattern of dust export over the NAO (Table 5.6). The zone of dust export is most intense between 10° N and 25° N, but it migrates seasonally. In JFM, the zone of maximum AI is between 5° N and 10° N. By AMJ, it is between 10° N and 20° N, whereas in JAS, it is between 15° N and 25° N. By OND, a southward retreat has begun, but dust export is relatively modest in amount. This seasonal pattern is comparable to that obtained from AVHRR aerosol optical thickness data (Swap et al. 1996) and from ship observations of haze made prior to the 1930s (McDonald 1938).

Specific sources for transatlantic dust plumes are not well known. They are perhaps most likely to be in the Mauritania–Mali area and further north in Western Sahara/Southern Morocco, although the clear seasonal signals found in dust concentrations on the western side of the Atlantic are not simply related to the seasonality of dust events recorded on the West African coast (Fig. 5.17). At Nouakchott, dust storms are a feature of the first six months of the year, before the annual rains; and hence it is unlikely that this station lies in the pathway of the strong summer flow that reaches Miami and Barbados. Further north, dust event frequencies at Nouadhibou and Dahkla are much less obviously seasonal, although the month of maximum dust activity at both stations is February. *Harmattan* dust blown from the Bodélé depression tends not to travel far over the Gulf of Guinea, as it is efficiently scavenged by rainfall associated with the Intertropical Convergence Zone, which typically descends no further than 5° N (Afeti and Resch 2000). The implication in various remote-sensing studies (e.g. Swap et al. 1996; Husar et al. 1997) that dust reaching South America may be from Bodélé is investigated further in Fig. 5.18, using data from the only period, 1977–1979, when regular monitoring of mineral dust was carried out at Cayenne (Prospero et al. 1981). The link with Bodélé dust is not confirmed by comparing the seasonality of mineral dust concentrations in surface level air at Cayenne with that of thick dust haze at Maiduguri in Northern Nigeria, a station directly within the *Harmattan* trajectory. A better, though still far from complete, link to potential sources is made with Nouakchott. However, the number of sites and length of data

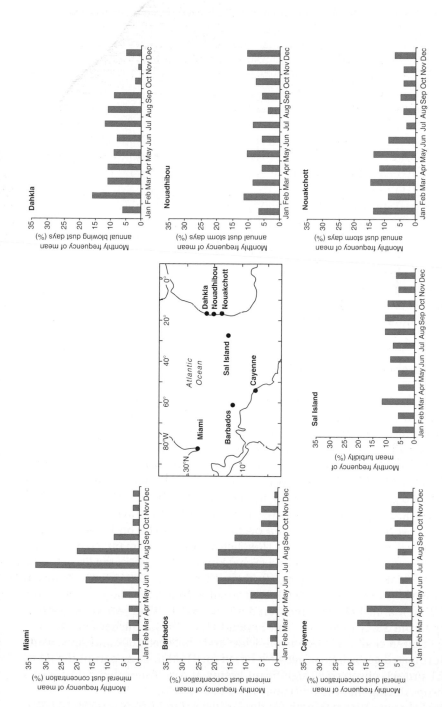

Fig. 5.17. Seasonality of dust events over the North Atlantic and west coast of Africa. Modified after Goudie and Middleton (2001, Fig. 3)

Table 5.6. Dust over the North Atlantic from 1999 TOMS data (percentage of days with AI >19)

Latitude (°N)	Season			
	JFM	AMJ	JAS	OND
45–50	0	0	0	0
40–45	0	0	0	0
35–40	0	0	0	0
30–35	0.44	0	2.24	0
25–30	0.22	8.64	9.78	0
20–25	0	24.58	29.52	0.68
15–20	0.68	33.1	29.12	2.26
10–15	3.44	30.0	8.7	0.68
5–10	11.9	11.7	1.32	0.44
0–5	3.44	0.68	0	0.22

considered are too limited for firm conclusions to be drawn and it may be that a combination of sources contributes to the Cayenne record.

5.6.2 European Trajectories

Saharan dust is often deposited over southern Europe in precipitation (Blanco et al. 2003) or in the dry form and has been reported since ancient times (Bücher and Lucas 1984). In southern Iberia, for example, the frequency of African dust outbreaks averages 16–19 episodes per year (Querol et al. 2004). In Mallorca, 253 Saharan dust rains were recorded between 1982 and 2003 (Fiol et al. 2005). In Italy, lidar observations in Naples suggest that the atmospheric aerosol load is influenced by Saharan dust about 15% of the time (Pisani et al. 2005). On 17 April 2005, a severe dust cloud enveloped Athens in Greece. Less frequently, deposition occurs further north, on the British Isles (Wheeler 1986), the Netherlands (Reiff et al. 1986), Germany (Jaeger et al. 1988; Littmann 1991b), Switzerland (Schwikowski et al. 1995), the French Alps (Aymoz et al. 2004), Hungary (Borbély-Kiss et al. 2004) and Northern Scandinavia (Franzen et al. 1994). Individual events can be large, such as the dustfall in March 1991, which covered at least 320 000 km², stretching across Europe from Sicily to Sweden and Finland (Burt 1991a; Bücher and Dessens 1992; Franzen et al. 1995). A Europe-wide study between 2000 and 2003, using a co-ordinated aerosol lidar network based on 21 stations as part of the EAR-LINET project, observed 90 significant events of free tropospheric dust layers in Europe during that period (Papayannis et al. 2005).

The Saharan source strength for dust transport to Europe was estimated at 80–120×10⁶ t year⁻¹ by D'Almeida (1986), based on sun photometer readings

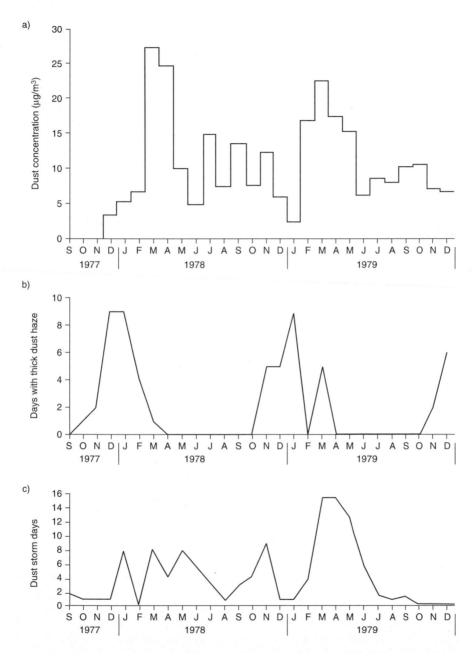

Fig. 5.18. a) Monthly mean mineral dust concentrations at Cayenne, French Guiana. Modified after Prospero et al. (1981). b) Monthly numbers of days with thick dust haze at Maiduguri, Nigeria. c) Monthly number of dust storm days at Nouakchott, Mauritania. Modified after Goudie and Middleton (2001, Fig. 4)

taken in the early 1980s, while the annual aeolian flux to the Western Mediterranean basin has been put at $3.9×10^6$ t by Löye-Pilot et al. (1986) who extrapolated from their monitoring of deposits at Corsica.

The seasonality of Saharan dust transport to Europe is shown for several parts of the continent in Fig. 5.19. An analysis of data from Hungary, Switzerland and Corsica indicates that the main period is from February to June, with a secondary maximum in the late autumn and early winter (Borbély-Kiss et al. 2004). For Britain, in the twentieth century (see Table 5.7), there is a similar bimodal distribution of activity, with a peak in March and another peak in September/October. By contrast, at Lannemezan in south-west France there is a clear peak in the summer months, with July having most outbreaks of Saharan dust and very few dust incursions occurring in the win-ter (Dessens and van Dinh 1990). The pattern in mainland Spain also has a peak of activity during the summer (i.e. May to August; Escudero et al. 2005), although generally it is more evenly distributed throughout the year (Rodriguez et al. 2001). This is also the case in Mallorca, where March to September is the prime season (Fiol et al. 2005). Similarly in Italy, May, June and July are the peak months for Saharan dust events (Rogora et al. 2004; Kischa et al. 2005). Interestingly, of 38 Saharan dustfall events noted in Britain in the twentieth century, not one occurred in the month of December. A dep-osition event that took place over Anglesey in North Wales in December 2003 (Perkins 2004) was the first on record for that month. Not surprisingly, Saharan dust falls on Britain have continued into the twenty-first century and include events on 25–26 February 2003, 17 September 2003, 12–13 February 2004, 1 April 2004, 1 April 2005, 30 April to 1 May 2005 and 31 August 2005.

A major source area for transport to Western Europe was identified by D'Almeida (1986) in southernmost Algeria, between Hoggar and Adrar des Iforhas. Another source, where material is particularly rich in palygorskite (Molinaroli 1996), is in Western Sahara–Southern Morocco. These sources have been confirmed by back-trajectory analysis for dust deposited over Northeastern Spain. Avila et al. (1997) traced deposition events back to three main areas: Western Sahara, the Moroccan Atlas and central Algeria. These sources have also been identified for transport to the British Isles (Tullet 1978; Wheeler 1986). A common trajectory for transport to Britain is over the Bay of Biscay, in mid-tropospheric winds skirting an anticyclone over Western Europe. A similar synoptic pressure distribution can deliver Saharan dust to the Iberian Peninsula (Rodriguez et al. 2001). Less commonly, dust is transported to Britain from Algerian sources across the Mediterranean and France in association with a low-pressure system centred over the Bay of Biscay (Wheeler 1986; Coudé-Gaussen et al. 1988). Again, such pressure sys-tems also deliver material to Spain. Algeria was found to be by far the most common source of Saharan dust deposited at Jungfraujoch in the Swiss Alps (Collaud Coen et al. 2004).

Transport to southern Europe occurs more frequently than to areas further north. A year of monitoring on Corsica, for example, revealed 20 dust events

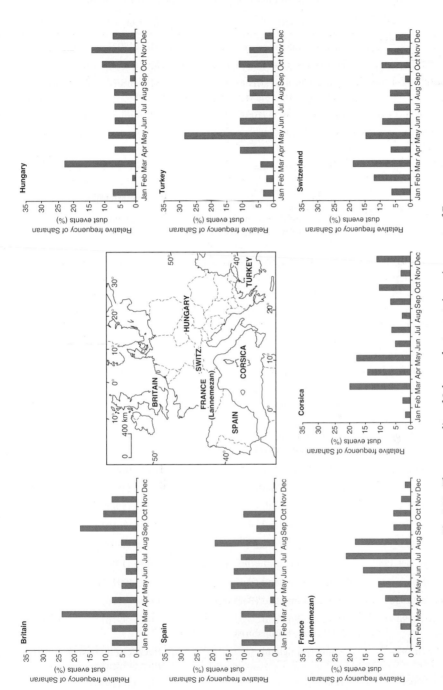

Fig. 5.19. The seasonality of Saharan dust events in various parts of Europe

Table 5.7. Known Saharan dust falls in the British Isles over the twentieth century

Date	Areas affected	References
9 March 1901	Central England	Mill (1902)
22–23 January 1902	South-west England	Mill (1902)
21–27 February 1903	Wales, south-west/central England, East Anglia	Mill and Lempfert (1904)
28 November 1930	English Channel coast	Alexander (1931)
1 July 1968	England and Wales	Pitty (1968), Stevenson (1969)
6 March 1977	Ireland, west Scotland	Tullett (1978), Bain and Tait (1977)
15 May 1979	Ireland	Tullett (1980)
28–29 November 1979	Ireland, North-west and central England, north Wales, south Scotland	Pringle and Bain (1981)
28–29 January 1981	North-west England, Northern Ireland	Richardson (1981), George (1981)
11 February 1982	South-eastern England	Thomas (1982), Moon(1982)
26–27 January 1983	Southern England, Somerset to Kent	Thomas (1983)
24 September 1983	Berkshire	Pike (1984)
29 September 1983	Northern Ireland	Tullett (1984)
22 April 1984	South Wales, Devon	Middleton (1986)
9 November 1984	Much of England and Wales, plus eastern Scotland	Thomas (1985), File (1986), Wheeler (1986), Cinderey (1987)
4 April 1985	South-east England (Kent, Cambridgeshire)	Thomas (1985)
5–6 March 1987	Southern England	Burt (1991b)
17–18 August 1987	England and Wales	Tullet (1988)
1 September 1987	Northern Ireland, Western Scotland	Tullet (1988)
17 September 1987	Southern England (Berkshire)	Burt (1991b)
6 October 1987	England and Wales	Tullet (1988)
26–27 October 1987	Eastern and Southern England	Smith (1988)
8 May 1988	Southern England (Berkshire)	Burt (1991b)
18 October 1988	Southern England (Berkshire)	Burt (1991b)
19 March 1990	Southern England (Berkshire, Hampshire)	Burt (1991b)
7–8 March 1991	Southern England	Bücher and Dessens (1992), Burt (1991a)
25 March 1991	Kent	Thomas (1993)
6 September 1991	Kent	Thomas (1993)

Table 5.7. Known Saharan dust falls in the British Isles over the twentieth century—cont'd

Date	Areas affected	References
11 October 1991	Southern England (Berkshire, Kent)	Burt (1992)
5 March 1992	Kent	Thomas (1993)
30 June 1992	Sheffield	Thomas (1993)
8 August 1992	Devon, Kent	Thomas (1993), Burt (1995)
17–18 September 1992	Devon, Berkshire, Kent	Knightley (1993), Thomas (1993)
16–17 March 1993	Berkshire	Burt (1995)
21 April 1994	Berkshire	Burt (1995)
24 September 1994	Central Southern England	Burt (1995)
14 February 1998	Ireland (Co. Mayo)	Sweeney (1998)
13 March 2000	Oxfordshire	Middleton et al. (2001)

(Bergametti et al. 1989) originating in three source areas: eastern Algeria/Tunisia/western Libya, Morocco/western Algeria and 'south of 30° N'.

The general synoptic pattern associated with dust transport from Africa towards Europe is discussed by Barkan et al. (2005). They suggest that the central importance lies in: (a) the strength and position of the trough emanating southward from the Icelandic low and (b) the eastern cell of the sub-tropical high. A deep, well developed trough near the Atlantic coasts of Europe and Africa, penetrating well to the south, and the strong eastern cell of the subtropical high situated to the north east of North Africa near the Mediterranean coast cause strong south western flows with the potential to carry dust northwards into the Mediterranean and on into Europe.

5.6.3 Eastern Mediterranean Trajectories

Dust transport from North Africa to the eastern Mediterranean (Fig. 5.20) occurs predominantly during the spring and is often associated with the eastward passage of frontal low-pressure systems – *Sharav* cyclones (Kubilay et al. 2003). These cyclones are principally formed by differential heating between relatively colder oceanic waters to the north and warmer landmasses to the south. Analysis of 23 heavy dust falls in Israel over a 20-year period suggests that the North African type is by far the most common (Ganor et al. 1991) and North Africa is also the main source of desert dust transported to Turkey (Mace et al. 2003; Kubilay et al. 2005). These storms are usually associated with a cold front with a significant downward-flowing jet stream and are often accompanied by rain (Alpert and Ganor 1993). Typically, the fronts are characterized by the presence of Saharan air at upper levels, above air from other source regions in the boundary layer, a situation confirmed by

Fig. 5.20. Suspended dust in the atmosphere over the eastern Mediterranean, 19 October 2002 (MODIS)

back-trajectory analysis conducted at 850 hPa and 500 hPa for air masses arriving at Erdemli in Turkey by Koçak et al. (2004). Dust from sources in the Middle East, by contrast, is more typically transported to the Mediterranean in the autumn (Dayan 1986; Kubilay et al. 2000, 2005) and tends to occur at higher altitudes (700 hPa and higher) than dust derived from North Africa.

Long-range transport of Saharan dust to the central Mediterranean basin is characterized by events lasting 2–4 days, compared to an average duration of just 1 day for events reaching the eastern Mediterranean from Arabia (Dayan et al. 1991). There is some seasonal variation in the source areas of dust reaching Israel, with Chad being the spring source, Egypt and the Red Sea the source in July/August and Libya in the autumn (Israelevich et al. 2003).

Central Algeria is the most frequent source area for Saharan dust reaching Israel (Ganor et al. 1991); and Ganor and Foner (1996) distinguish between material commonly transported from sources in the Hoggar Massif and the Tibesti mountains in Northern Chad, the latter also picking up material from the Western and Sinai Deserts.

5.7 Middle East

Dust storms are important phenomena over large tracts of the arid and semi-arid regions of the Middle East (Middleton 1986a; Kutiel and Furman 2003; Leon and Le Grand 2003). Indeed, Arabia was identified by Idso (1976) as one of the five major world regions where dust storm generation is especially intense. A number of dust-bearing winds have been identified (Table 5.8).

A preliminary analysis of the distribution and seasonality of dust storms is provided by Middleton (1986a), who, on the basis of the analysis of meteorological data established that southern Iraq (Al-Najim 1975) and Kuwait (see also Abdulaziz 1994) had the highest number of dust episodes (Fig. 5.21). At stations in Qatar, Kuwait and Iraq, dust activity is at its highest from April to August. A thorough analysis of the situation in Kuwait is provided by Safar (1985). Subsequently, on the basis of the study of aerosol geochemistry over the Arabian Sea, Pease et al. (1998) suggested that the Wahiba Sands area of Oman is also a major dust source region. Dust storms are common in the Saudi Arabian city of Riyadh (Modaish 1997), where 41 days with a visibility less than 1.6 km occur each year. There is also considerable dust storm

Table 5.8. Dust-bearing winds of the Middle East

Name	Area affected	Season	Direction	Meteorological conditions
Shamal (north)	Mesopotamia and Arabian Gulf	Feb./Oct.	N	Blows between Azores high and Indian monsoon low
Sad-ou-bist bad (wind of 120 days)	South-eastern Iran (especially Seistan)	May/Sept.	N–NW	Circulates around main low of Indian monsoon
Belat	South-eastern Arabia especially between Ras Sajar and Masira Island	mid-Dec./ mid-March	N–NW	
Simoom (poison wind)	Kuwait	Summer	NW	
Khamsin (fifty)[a]	Egypt	Spring and winter	Predominantly S	Local wind caused by particular air masses drawn into region by passage of a cyclone and its associated fronts
Sharav	Israel	April–June	SW–S or SE	Khamsin-type
Shlour	Syria and Lebanon	Spring and winter	S–SW	
Shargi	Ira	Spring	SE	

[a] Variously taken to refer to the average duration of wind (50 h), its annual frequency (50 times) and its season of maximum onset (the 50 days either side of spring solstice). From: Middleton (1986a Table 1)

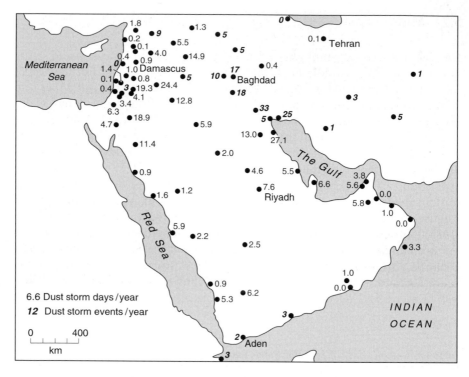

Fig. 5.21. The distribution of dust storms (visibility less than 1 km) in the Middle East as determined from meteorological data. Modified after Middleton (1986a, Fig. 4)

activity in the Negev (Offer and Goossens 2001) and studies of dust deposition have also been made over the Dead Sea (Singer et al. 2003). Leon and Le Grand (2003), using IDDI from Meteosat, give a regional picture for the whole of the North Indian Ocean region.

Dust deposition contributes to the formation of loess deposits at various locations in the Middle East, including the Negev Desert, Yemen, Saudi Arabia, Syria, Bahrain and the United Arab Emirates (Goudie et al. 2000). It also contributes to sedimentation in the Arabian Gulf (Sugden 1963; Foda et al. 1985), the Eastern Mediterranean (Kubilay et al. 2000), the Red Sea and the Arabian Sea (Stewart et al. 1965; Prins et al. 2000). Dust is also a major hazard for engineering structures and for air quality in the region (Jones 2001).

Some of the dust in the Middle East is locally derived (Fig. 5.22), but significant quantities come into the Levantine Basin (Krom et al. 1999), Turkey (Güllü et al. 2005) and Israel (Ganor and Foner 1996) from the Sahara. Figure 5.23 shows simplified maps of seasonal changes in the AI determined by TOMS. In January, February and March (JFM), the area with AI >6 lies largely in the south and east of the Arabian Peninsula south of latitude 32° N. The most intense area of activity, with a small stretch where AI >15, is on the Oman–Saudi Arabia border at ca. 20° N. In April, May and June (AMJ),

Fig. 5.22. A dust storm at Jazirat al Hamra, near Ras al Khaimah, United Arab Emirates (from ASG)

the situation is dramatically changed, with much of the Middle East south of ca. 37° N experiencing AI values >5. An area of AI >5 has also developed on the east side of the Caspian Sea and the same is true of Iran. The area with AI >15 has expanded to include a large swathe of interior Arabia and part of the Makran coast of Iran. The Oman–Saudi border region continues to be the most developed area of dust, but the AI values now exceed 25. In July, August, September (JAS), the situation is broadly similar to that in AMJ. However, by October, November, December (OND), the area with AI >6 has shrunk very noticeably, being restricted to southern and eastern Arabia. There is only a very small area, the Oman–Saudi border region, where AI >15. The contraction of the area with high AI values in the winter season (JFM, OND) is related in all probability to the occurrence of rainfall in the northern part of the region during the winter months. However, this is also the season when cyclonic activity is most likely to occur; and Offer and Goossens (2001, Fig. 21) suggest that the peak of dust storm activity in the Negev in February may be related to this cause. The intensification of dust storm activity in the southern part of the region during the summer months (AMJ, JAS) may be related to a variety of factors, including dust inputs from the Sahara, for these are the months when the northern part of 'the Saharan dust machine' is most active (Goudie and Middleton 2001). It is also a time of intense atmospheric instability because of the extreme surface temperatures that are achieved. In addition, it is a time when strong north-westerly winds – the *Shamal* – occur. In Arabia as a whole (Table 5.9), OND has the lowest wind velocities,

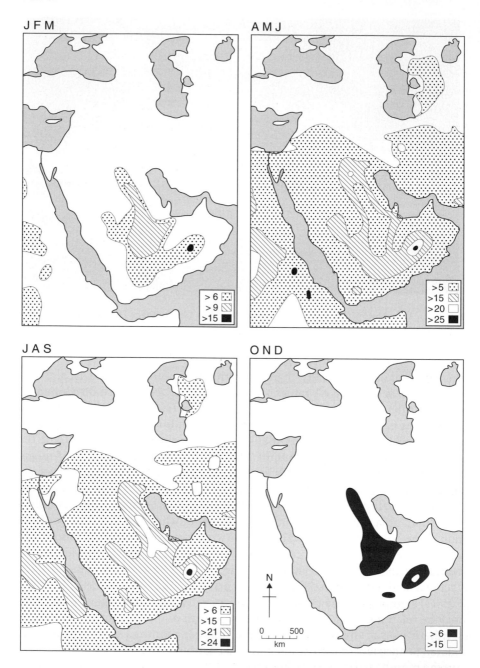

Fig. 5.23. The seasonal pattern of dust storm activity in the Middle East, derived from TOMS. The values are long-term values of the aerosol index (AI). Modified after Goudie and Middleton (2002, Fig. 1)

Table 5.9. Seasonality of Arabian Wind (mean wind speed; m s^{-1})

Location	J	F	M	A	M	J	J	A	S	O	N	D
Dhahran	0.9	0.9	1.0	1.1	1.0	1.2	0.9	0.9	0.9	0.8	0.7	0.7
Jeddah	0.9	1.0	0.9	0.8	0.8	0.8	0.7	0.9	0.8	0.6	0.5	0.5
Madinah	0.6	0.5	0.6	0.6	0.6	0.6	0.7	0.6	0.5	0.5	0.5	0.4
Riyadh	0.6	0.6	0.5	0.5	0.6	0.6	0.6	0.4	0.3	0.3	0.3	0.3
Taif	0.7	0.8	0.8	0.7	0.6	0.8	0.1	0.9	0.6	0.6	0.5	0.4
Bahrain	5.2	5.4	5.0	4.6	4.9	5.8	4.3	4.3	3.7	4.0	4.3	5.0
Doha	4.5	4.7	4.9	4.8	5.0	5.5	4.4	4.4	3.5	3.8	4.0	4.2
Abu Dhabi	3.8	4.3	4.6	4.0	4.1	4.1	3.9	4.0	3.6	3.2	3.1	3.5
Dubai	3.0	3.6	3.6	3.6	3.8	3.9	3.6	3.6	3.3	3.0	2.8	3.0
R.A.K.	2.2	2.3	2.6	2.8	2.9	2.8	2.8	2.8	2.3	2.0	2.0	2.0
Sharjah	3.3	3.5	3.6	3.7	3.8	3.7	3.6	3.5	3.2	2.9	3.0	3.0
Amman	3.2	3.6	3.6	3.6	3.5	3.9	4.1	3.6	2.7	2.3	2.5	2.9
Deir-Alla	2.2	1.9	1.8	1.9	1.8	1.5	1.5	1.5	1.5	1.6	2.2	2.3
Irbid	6.9	7.2	7.1	6.9	7.0	8.8	9.7	8.8	6.7	4.9	5.4	6.0
Kuwait	3.7	4.1	4.5	4.5	4.7	5.8	5.6	4.8	3.7	3.5	3.5	3.5
Monthly mean	2.78	2.96	3.01	2.93	3.13	3.32	3.17	3.17	2.49	2.27	2.35	2.51
Quarterly mean			2.92			3.13			2.94			2.37

whereas the highest velocities occur in MJJA. This seasonal pattern is confirmed by visibility data for Masirah Island off Oman. Mean monthly visibility is at its lowest in MJJA. Likewise, data from the ground-based Aerosol robotic network (AERONET) show that the maximum dust aerosol loading in Bahrain occurs in the March–July period (Smirnov et al. 2002).

5.7.1   The Spatial Pattern of Dust in the Middle East from TOMS

The mean annual AI values for Arabia and neighbouring areas are mapped in Fig. 5.24. It is clear that substantial dust loadings occur over much of the Arabian Peninsula and that the values are comparable to those obtained over large tracts of the eastern Sahara. By looking at the AI values and their areal extent, it is possible to gain an indication of the strength of dust loadings over Arabia in comparison with other desert areas. As Table 4.2 indicates, the dust source on the Oman/Saudi Arabia border is the third strongest in the world, only being exceeded by the western and central Saharan sources.

There is a clear tendency for the highest AI values to occur in the south and eastern Arabia. One intense area is on the borders of Oman and Saudi Arabia centred at ca. 19° N and 54° E. This is a very dry, low-lying area fed by a series

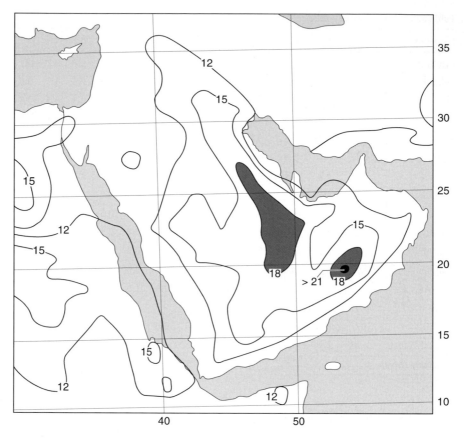

Fig. 5.24. The annual pattern of dust storm activity in the Middle East, derived from TOMS AI values. Modified after Goudie and Middleton (2002, Fig. 2)

of ephemeral wadis that have their sources in the mountain rim of Yemen and Oman. It also includes a large area of closed drainage with numerous playas, including the Umm and Samim and the Sabkhat Aba ar rus. Glennie and Singhvi (2002) show the extent of the 100-m closed contour in their Fig. 1 and outline it in their geomorphological map of SE Arabia as a 'deflation' plain. The other dust 'hot spot', which is larger in extent but less intense, is in eastern Saudi Arabia to the north of the great Rub Al Khali sand sea. The mountainous rims of Arabia (Fig. 5.25) and the more humid areas of the Middle East are not major dust source regions. Dust storms are most prevalent where the mean annual precipitation is less than 100 mm and mean annual potential evapotranspiration is over 1140 mm. The concentration of dust storms in areas where the mean annual rainfall is less than 100 mm confirms a picture that emerges from the Sahara (Middleton and Goudie 2001) but is at variance with the suggestion of Goudie (1983), based on analysis of meteorological observations, that the driest areas are not as important for dust storm generation as those with rather higher amounts.

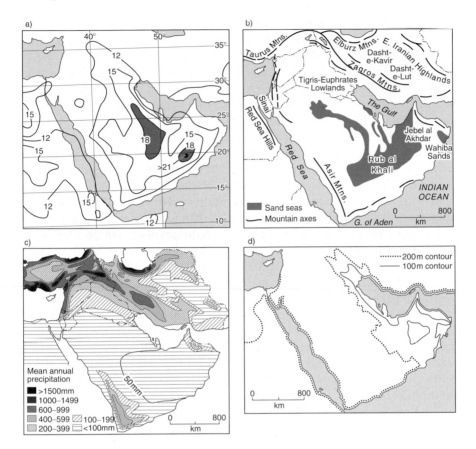

Fig. 5.25. a) The annual pattern of dust storm activity in the Middle East in relation to: b) the topography, c) the precipitation of the region, and d) areas below 200 m. Modified after Goudie and Middleton (2002, Fig. 3)

When one compares the TOMS picture with that obtained from ground meteorological observations (Fig. 5.21), certain major differences are apparent. However, it has to be remembered that such differences are partly a result of the paucity of meteorological stations in some parts of inner Arabia, most notably in the south-east quarter where TOMS shows the highest AI values. Nevertheless, some of the highest dust storm occurrences recorded by ground stations occur in the vicinity of Kuwait, Baghdad and Basra (Middleton 1986a), yet this is not an area identified by TOMS as having very high dust loadings, though they are still appreciable.

5.7.2 Dust Transport from the Sahara to the Middle East

Although much dust is raised locally over the Middle East, it is apparent that substantial amounts of dust come from the Sahara. The type of synoptic situation responsible is the passage of an advancing cold front and the associated strong surface winds ahead of it penetrating south-eastwards from the Mediterranean Sea deep into the northern Sahara and Libyan Desert (Michaelides et al. 1999). This can be illustrated by the situation in mid-March, 1998, when a major dust event caused ports and airports to be closed, created breathing problems for inhabitants of Amman and led to fatal motoring accidents in Egypt and Jordan. Mean visibility in Amman was reduced to 4.2 km. A large, deep depression moved eastwards from North Africa and then deepened further over the Middle East as it encountered cold polar air pushing across Turkey. The progress of this system can be traced by looking at the daily TOMS maps for the period 14–20 March (Fig. 5.26). The sequence starts with an area in eastern Algeria, southern Tunisia and north-western Libya generating AI values greater than 26. The following day, it has moved across Libya into Egypt and the eastern Mediterranean. On 16 March, the main area with high AI values covers Cyprus and the Levant. Mean visibility at Amman airport was reduced to 3.2 km. On 17 March, the area with high AI values has broken down into a group of small, deep clusters and by 19 March, most of the area has AI values less than 10.

Another example is provided by a TOMS sequence for late March 1999 (Fig. 5.27). On 27 March, a deep system developed over northern Libya, and achieved AI values that exceeded 27. The system moved rapidly eastwards across northern Egypt on 28 March and to northern Saudi Arabia on 29 March. It then weakened as it moved down the Gulf, reaching Bahrain and Qatar on 30 March.

The Red Sea shore of Africa can be another major source over western Saudi Arabia, particularly the plain of the ephemeral Baraka river and the Tokar delta. The TOMS image for 24 March (Fig. 5.28) shows the presence of an area of relatively high AI values heading across the Red Sea into Saudi Arabia in the vicinity of Jiddah, Makkah and At Ta'if. Similarly, in late June 2000, a series of dust pulses travelled from the northern Sudan across the Red Sea into Saudi Arabia and could be traced both on TOMS (Fig. 5.29) and on AVHRR imagery.

Yet another example of the movement of a dust storm from the northern Sahara to the Middle East is provided by the events of April 2000 (Fig. 5.30). On 18 April, TOMS showed a large area of dust over southern Libya and far western Egypt. On the following day this reached eastern Egypt, Israel and Lebanon, causing the closure of the port of Alexandria and the cessation of flights to Aswan. The mean visibility at Cairo airport was reduced to 4.6 km and that at Luxor to 3.7 km. In Israel, mean visibility in Beersheva was reduced to 2.6 km. Limassol in Cyprus was also badly affected, as were flights in southern Turkey.

In March 2002, a large dust storm blew from north eastern Africa across to Iran (Fig. 5.31). On 19 March, the system had intensified over Israel and Palestine and by 20 March it had moved eastwards to the Tigris–Euphrates valley and the north-east of Iran. On 21 March, dust remained over southern Iraq.

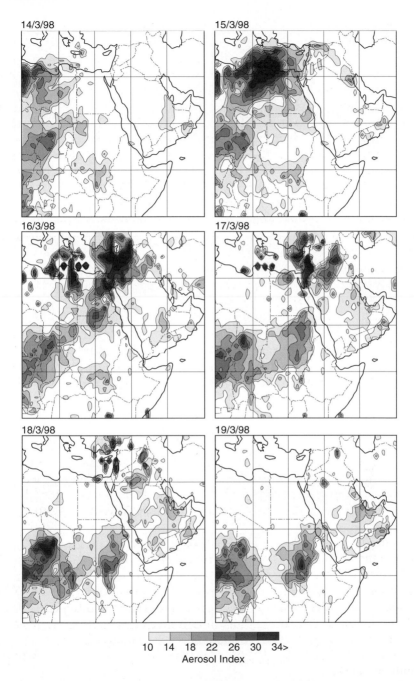

Fig. 5.26. The passage of dust systems from North Africa to the Middle East, mid-March 1998, based on TOMS AI values. Modified after Goudie and Middleton (2002, Fig. 5)

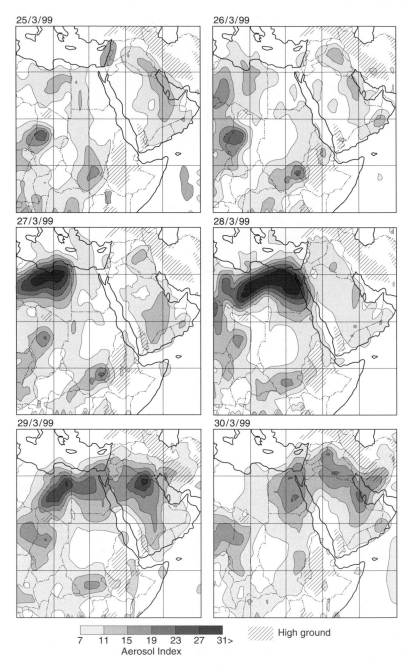

Fig. 5.27. The TOMS AI sequence for late March 1999. Modified after Goudie and Middleton (2002, Fig. 6)

24/3/00

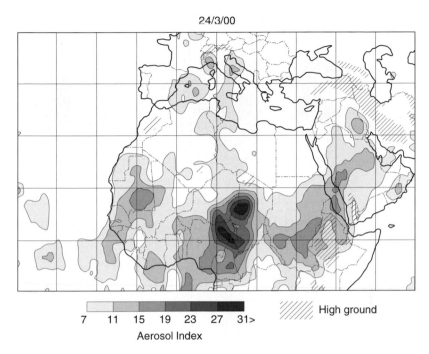

High ground

7 11 15 19 23 27 31>

Aerosol Index

Fig. 5.28. The TOMS AI sequence for 24 March 2000. Modified after Goudie and Middleton (2002, Fig. 7)

5.8 South West Asia

Dust storms are widespread in the northern part of the Indian sub-continent and neighbouring areas (Léon and Le Grand 2003; El-Askary et al. 2005). Middleton (1986b) used ground station observations to examine the frequency and seasonality of dust storms in south-west Asia. Figure 5.32 is his map of dust storms in the region. It shows that the highest frequencies occur at the convergence of the common borders between Iran, Pakistan and Afghanistan. Other high-frequency areas occur on the Arabian Sea coast of Iran (Makran) and across the Indus Plains of Pakistan into north-west India (Hussain et al. 2005) and the Indo-Gangetic basin (Dey et al. 2004). Littmann (1991a) also mapped the frequency of Asian dust storms and examined some of the climatic factors that control their seasonal occurrence. The geochemistry of the dust aerosols in the vicinity of the Thar Desert are discussed by Yadav and Rajamani (2004).

Multiple dust sources are discernible on the annual mean map of TOMS data (Fig. 5.34). These sources are broadly concurrent with those mapped by Middleton (1986b; Fig. 5.32). Figure 5.33 shows four major source areas with

A.S. Goudie and N.J. Middleton

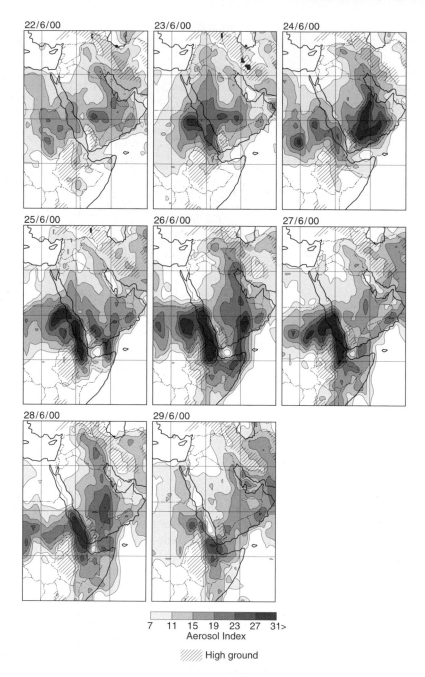

Fig. 5.29. The TOMS AI sequence for late June 2000. Modified after Goudie and Middleton (2002, Fig. 8)

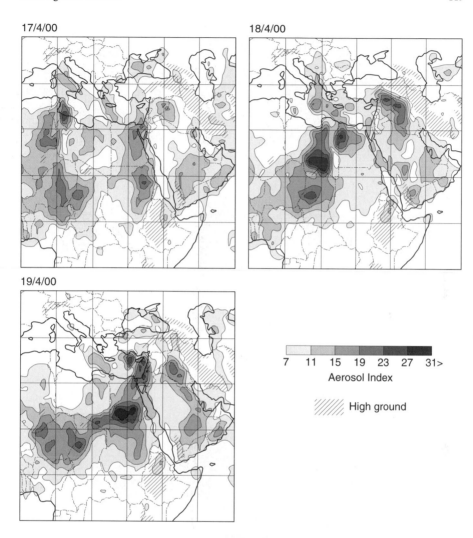

Fig. 5.30. The TOMS AI sequence for mid-April 2000. Modified after Goudie and Middleton (2002, Fig. 9)

AI values of >8: (a) the Makran coastal zone, stretching from south-eastern Iran into neighbouring Pakistan, (b) a broad area of central Pakistan, (c) an area at the convergence of the borders of Iran, Afghanistan and Pakistan that comprises the Seistan Basin (Fig.5.35), the Registan sand sea and north-western Baluchistan and (d) an area approximately coincident with the Indus delta. A broad "tongue" of dust-raising activity stretching south westwards down the alluvium of the Gangetic plain is also clearly defined on both maps. Some of the dust loading in this latter area may come from as far away as the Arabian Gulf (Dey et al. 2004) or the Sahara (El-Askary et al. 2005).

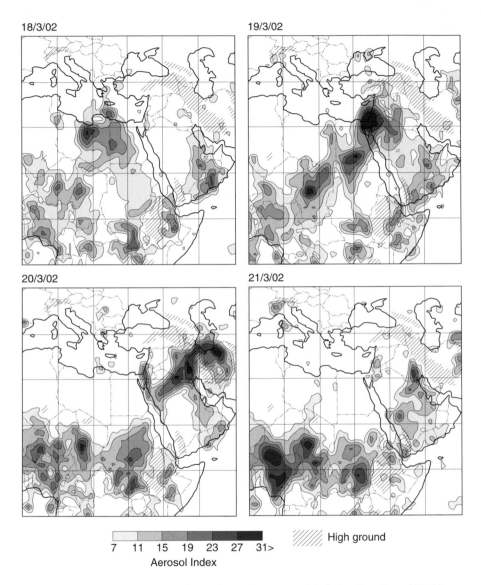

Fig. 5.31. The TOMS AI sequence for mid March 2002. Modified after Goudie and Middleton (2002, Fig. 10)

Coastal Baluchistan/Makran appears as the most active source area according to the TOMS data, whereas Middleton's (1986b) map (Fig. 5.32) shows the Seistan Basin area to have the most frequent dust storm activity. Middleton does not record the Indus Delta as a significant area for dust storm activity, having fewer than five dust storms a year. However, Middleton highlights the plains of Afghan Turkestan as an area where annual dust storm

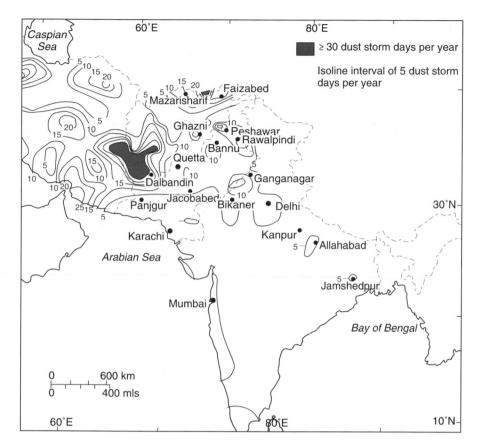

Fig. 5.32. The number of dust storm days per year in South Asia, based on ground observations. Modified after Middleton (1986b)

frequency exceeds 30 and two areas in Iran (around Yazd in the centre and along the border with Turkmenistan) as having 20 or more dust storm days annually. None of these areas appears significant according to the TOMS data.

The Makran is a hyperarid area of late-Quaternary uplift (Vita-Finzi 1981; Reyss et al. 1998). Material is supplied to the coastal strip from the mountains inland; and silt-sized material blown from ephemeral rivers and alluvial fans southward over the Arabian Sea (Fig. 5.36) dominates near-shore sediments (Mohsin et al. 1989).

The Iran/Afghanistan/Pakistan border area is known as the Dasht-i-Margo. Dust sources are found in lowland parts of this mountainous region, including the Seistan Basin. This is a huge closed depression, around 450 km across, so that by analogy with areas like Bodélé, Taklamakan and Eyre, it is perhaps not surprising that it is a very active dust source. Sediments available for deflation are fed into the basin from the surrounding mountains. Specific source areas are likely to be alluvial fans and ephemeral lakes. Indeed,

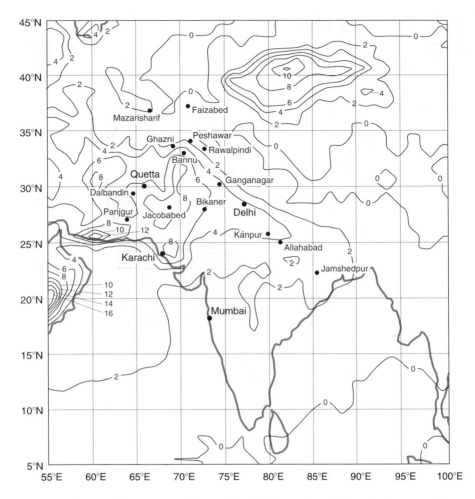

Fig. 5.33. The annual TOMS mean for South Asia. The scale on this and subsequent figures shows the aerosol index (AI). Modified after Goudie and Middleton (2000, Fig. 2)

MODIS images of the area show that the bed of Lake Hamun and the large deltaic fan of the Helmand River, which flows into it, are repeated sources of dust storms. This is probably caused in part by desiccation of the area brought about by diversion of upstream water for irrigation use (see www.unep.org/governingbodies/gc22/document/afghanistan4.pdf) and by extreme droughts in recent years. Dense plumes of dust originating from the dried lake beds and from the delta of the Helmand are transported by high-velocity winds coming from the north and funnelled by gaps in the high mountains. The famous 'wind of 120 days' was discussed by early travellers to the region. McMahon (1906, p. 224), for example, wrote: " It sets in at the end of May or the middle of June, and blows with appalling violence, and

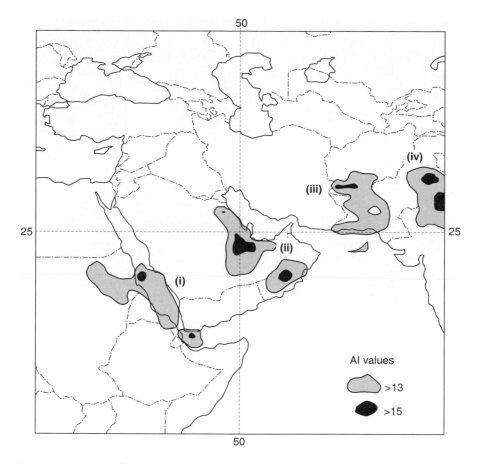

Fig. 5.34. Dust storm hotspots in the north-west Indian Ocean region 1998–2002 from TOMS

with little or no cessation, till about the end of September. It always blows from one direction, a little west of north, and reaches a velocity over 70 miles an hour. It creates a pandemonium of noise, sand and dust". He noted that it left old irrigation canal beds, which are more resistant than surrounding sediment, standing above the level of the adjacent land, and that there were some wind scour features around 6 m deep.

5.8.1   The Seasonal Cycle of Dust from Ground Observations

Table 5.10 presents data on dust storm seasonality for a range of climatological stations in Afghanistan, India and Pakistan. There is some variability in the month with maximum dust activity, with all months between March and October having at least one station where this occurs. Equally, no stations

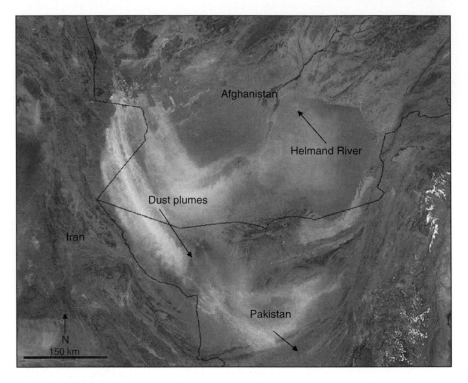

Fig. 5.35. A MODIS view of a dust storm blowing off the Seistan Basin of south-west Asia, 17 August 2004

have maximum monthly frequencies between November and February. When one takes the mean for all 17 stations used, the dustiest period covers May and June, when just over 40% of all dust storms occur. This is the pre-monsoon season (Hussain et al. 2005). Only 7.8% of dust storm activity occurs between November and February.

5.8.2 The Seasonal Cycle of Dust from TOMS

In January, February and March, the area with reasonably high AI values is small, and the highest AI values are less than 8 (Fig. 5.37). There is one zone located on the Makran coast of Iran and another in the lower Indus plain where AI values lie between 6 and 8. By March, April and May (Fig. 5.38), the situation is transformed and there is now a large belt from Iran across to north west India where AI values exceed 10. There is a strong zone of dust activity along the Makran coast where AI values exceed 14 and another along the Ganges Plain where values exceed 12. In April, May and June, just before the break of the south west monsoon (Fig. 5.39) the AI values reach their

Table 5.10. Seasonality of dust storms (frequency as % by month) in Afghanistan, Pakistan and India. Months with largest frequency of dust storms are shown in bold

	J	F	M	A	M	J	J	A	S	O	N	D	Ave. no. per year
Afghanistan													
Bust	4.7	9.5	10.4	13.7	10.4	8.5	10.4	**14.2**	5.7	4.3	3.3	4.7	30.1
Ghazni	0.0	0.0	2.2	**20.0**	13.3	11.1	13.3	14.8	8.1	10.3	5.2	1.5	19.3
Mazarisharif	0.8	0.8	4.8	4.8	4.0	15.9	15.1	13.5	7.1	**23.0**	8.7	1.6	18.7
Faizabed	0.0	0.0	1.4	7.1	4.3	14.3	20.0	**22.9**	8.6	17.1	4.3	0.0	17.5
Pakistan													
Bannu	0.0	1.2	5.9	4.7	19.6	15.7	**23.5**	15.7	11.8	2.0	0.0	0.0	25.5
Dalbandin	3.5	7.0	14.0	14.0	14.0	14.0	**17.5**	7.0	4.2	2.8	1.0	1.0	28.6
Jacobabed	1.1	0.0	16.3	12.0	18.5	12.0	**21.7**	12.0	4.3	0.0	0.0	2.2	9.2
Panjgur	3.4	17.2	**31.0**	3.4	6.9	17.2	13.8	3.4	0.0	3.4	0.0	0.0	3.6
Peshawar	0	7.4	1.5	3.7	**22.2**	14.8	22.2	14.8	12.6	6.7	7.4	0.0	13.5
Quetta	0.0	1.8	7.1	5.4	12.5	17.9	5.4	12.5	**19.6**	16.1	0	1.8	5.6
Rawalpindi	0.0	1.4	4.3	14.2	**21.3**	21.3	14.2	9.9	7.1	5.7	0.7	0.0	14.1
India													
Ganganagar	8.9	0.0	11.1	0.0	**33.3**	24.4	13.3	8.9	0.0	0.0	0.0	0.0	17.0
New Delhi	0.0	0.0	10.0	10.0	**40.0**	35.0	3.3	0.0	0.0	1.7	0.0	0.0	8.0
Kanpur	4.4	2.2	8.9	13.3	**44.4**	30.0	0.0	0.0	0.0	4.4	2.0	0.0	5.0
Jamshedpur	0.0	0.0	7.1	23.8	**50.0**	16.7	2.4	0.0	0.0	0.0	0.0	0.0	6.0
Bikaner	1.7	6.7	9.5	11.2	16.8	**27.9**	11.2	7.3	3.4	3.4	0.0	1.1	17.9
Allahabad	0.0	5.9	3.9	13.7	**39.2**	29.4	5.9	0.0	0.0	2.0	0.0	0.0	5.1
Mean	1.68	3.59	8.79	10.29	21.81	18.60	12.54	9.23	5.44	6.05	1.72	0.81	–

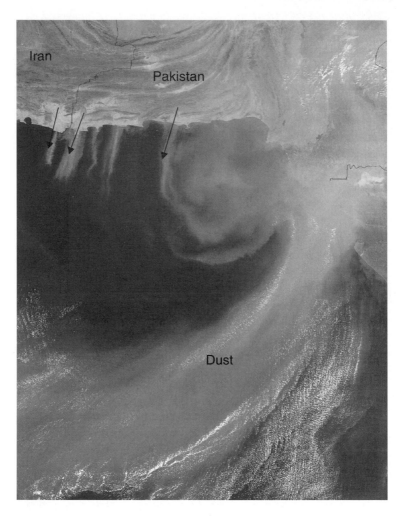

Fig. 5.36. Plumes of dust from the Makran coast of Iran and Pakistan are captured in this MODIS image on 14 December 2003

annual peaks. There is a large expanse of country where they are greater than 15 and two locations (the Makran coast and the Sibi Plain of Baluchistan), where values exceed 18. By July, August and September (Fig. 5.40), the spread and intensity of the zone of high dust loadings have contracted. The Ganges Plain is no longer significant and AI values in the Indus Plain are less than 18. The Makran, however, continues to be important, with some AI values greater than that figure. In October, November and December (Fig. 5.41), the Indian region is at its least dusty condition during the annual cycle. AI values are low throughout the region, and do not exceed 6. The two hot spots – the Makran coast and the southern Indus valley – are, however, evident.

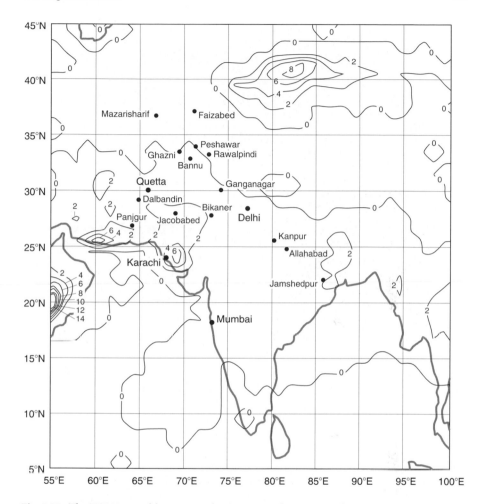

Fig. 5.37. The TOMS monthly mean AI for January, February, March

5.8.3 Climatic Relationships to Dust Seasonality in South Asia

The explanation for the extreme seasonal variation in dust activity revealed both by ground observations and by TOMS lies with various climatic factors. The predominant factor is the seasonality of rainfall, which in turn controls soil moisture content (cohesiveness) and vegetation cover. The south-west summer monsoon brings a maximum of precipitation to the south and east of the dry zone, with July and August being especially wet. In the north and west of the region (e.g. in Baluchistan and the North-West Frontier of Pakistan), the rainfall maximum may be in late winter. The contraction of the area of dust activity from the Ganges Plain and elsewhere in July to September (Fig. 5.42) can be explained by the high number of rainy days at that time.

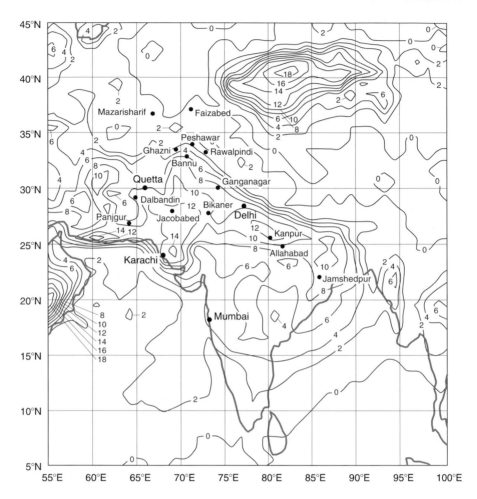

Fig. 5.38. The TOMS monthly mean AI for March, April, May. Modified after Goudie and Middleton (2000, Fig. 3)

Another important control of dust storm activity is the occurrence of thunderstorms, for these are one of the main factors that can generate dust from the ground surface. Although for the area as a whole (Table 5.11) the highest frequency of thunderstorms is during the wet months of July and August, there is also substantial activity in May and June, prior to major precipitation occurring with the onset of the southwest monsoon.

Wind activity, a crude measure of which is wind velocity (Table 5.11), is closely related to thunderstorm frequency, with the highest mean wind velocities occurring in early summer. Also important are pressure conditions. The easterly movement of 'western disturbances', low-pressure zones either at the surface or in the upper westerly wind regime north of the subtropical high pressure belt, are responsible for two distinct synoptic situations that

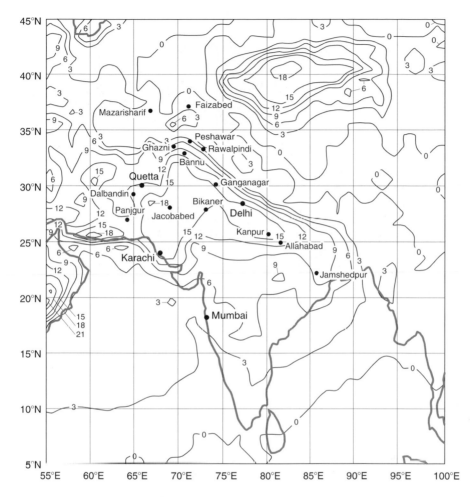

Fig. 5.39. The TOMS monthly mean AI for April, May, June. Modified after Goudie and Middleton (2000, Fig. 4)

cause dust-raising over much of the area. These troughs move across Iran and Turkmenistan to affect the Indian subcontinent north of 30° N. Weak circulations, called induced lows, may simultaneously develop over central parts of Pakistan and Rajasthan and move east-north-eastwards (Rao 1981).The two dust-raising situations commonly caused by these lows are the creation of a steep pressure gradient, where strong winds may cause deflation from parched soils, and the creation of an area prone to thunderstorm generation, where dust is mobilized by the dry thunderstorm downdraft. Dry, dust-raising thunderstorms are meso-scale phenomena, typically lasting less than an hour at any one spot, as the thunderstorm system moves with typical speeds of 60 km h^{-1}. These storms are most common in north-west India,

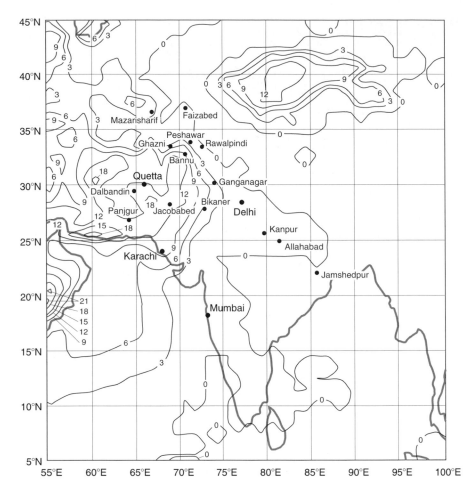

Fig. 5.40. The TOMS monthly mean AI for July, August, September. Modified after Goudie and Middleton (2000, fig. 5)

where they are known as *Andhi,* the majority of which occur during the pre-monsoon hot season (April–June).

The pressure-gradient dust storms are synoptic scale features that can raise dust over large areas throughout Pakistan and north-western India, often continuing for several days (Middleton 1989). Once raised, dust can then remain in the atmosphere for several days, being generally transported towards the east or north-east in the pressure-gradient winds. Such material, when transported in lighter winds, creates dust haze conditions known as *Loo.* This is typically experienced to the east and north-east of Rajasthan, in Delhi and on the Ganges plain as far east as Bihar.

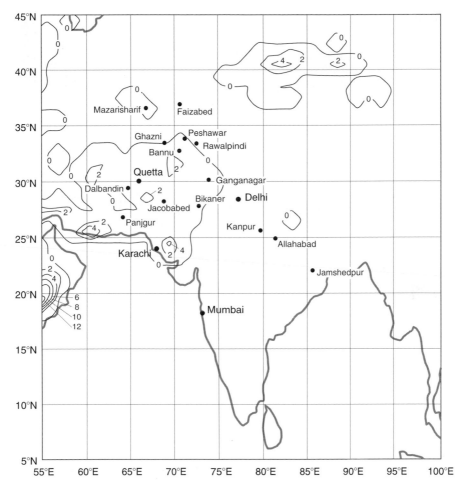

Fig. 5.41. The TOMS monthly mean AI for October, November, December. Modified after Goudie and Middleton (2000, fig. 6)

To the north and east of Rajasthan, the Loo's role becomes less important and that of the Andhis more important. Joseph et al. (1980) state that most of the dust storms occurring at New Delhi are of the Andhi type, a situation exemplified in Fig. 5.42a, which shows that the peak dust storm months of May and June correspond to a high frequency of thunderstorms. Although thunderstorm frequency rises further in July and August at New Delhi, these months are also associated with high monsoon rainfall totals. Maximum dust storm frequencies at Ganganagar are also experienced in May and June (Fig. 5.42b) but these are not months of elevated thunderstorm frequency. Dust-raising here is more closely associated with the pressure-gradient winds.

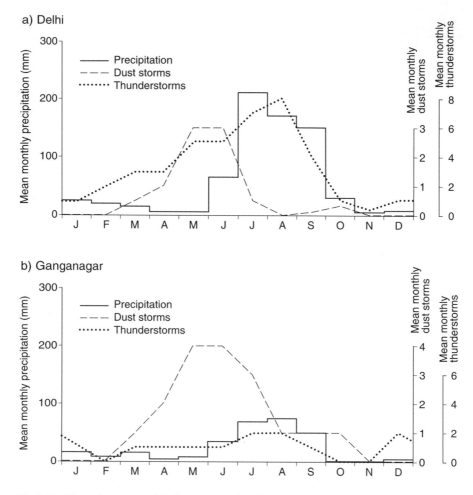

Fig. 5.42. Plots of mean monthly dust storms, thunderstorms and rainfall for: a) New Delhi and b) Ganganagar. Modified after Goudie and Middleton (2000, Fig. 7)

To summarize, in the winter, although it is dry over most of the region, dust storm activity is low. This is because of high-pressure conditions, a lack of thunderstorm activity and the absence of strong winds. In the pre-monsoon season, conditions are still dry, but wind velocities and thunderstorm activity increase. This is a time when strong heating of the landmass generates unstable conditions and convective low-pressure systems, generating maximum dust activity. The onset of the monsoonal period in July leads to a sharp decrease in dust activity. Soil water storage and the persistence of a vegetation cover ensures that dust storm activity remains at low levels into the winter months.

Table 5.11. Monthly frequency of dust storms, thunderstorms and mean wind speeds for the desert region of the Indian sub-continent

	J	F	M	A	M	J	J	A	S	O	N	D
Dust storms (frequency as % by month)	1.68	3.59	8.79	10.29	21.81	18.6	12.54	9.23	5.44	6.05	1.72	0.81
Thunderstorms (frequency as % by month)	2.35	2.55	7.85	9.31	10.35	14.07	18.64	16.14	10.84	4.46	1.02	2.4
Wind speeds (mean velocity m s^{-1})	1.6	1.8	2.1	2.3	2.8	3.2	3.1	2.6	2.2	1.5	1.3	1.3

5.9   Central Asia and the Former USSR

In the southern portions of the former Soviet Union there is a large zone where the number of dust storms exceeds 40 per year (Klimenko and Moskaleva 1979) and some locations where there are more than 80, one of the highest occurrences in the world (Fig. 5.43). May to August is the period with greatest activity; and Kazakhstan was identified by Zakharov (1966) as having the greatest frequency of occurrence. Human activities have caused dust

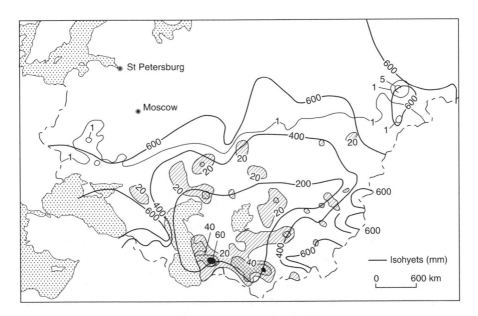

Fig. 5.43. The distribution of dust storms in the former Soviet Union. Based on the work of Kes and Fedorovich, in Goudie (1983a, Fig. 5)

storm frequencies to be raised both by the extension of cultivation, particularly during the ploughing up of pastures associated with the Virgin Lands Scheme of the 1950s and as a result of the desiccation of the Aral Sea. The 'white', saline dust from the former seabed of the Aral region is highly toxic (O'Hara et al. 2000) and links have been suggested between this atmospheric dust and poor human health in the region (Wiggs et al. 2003; see also Chapter 7). Sixty per cent of storms in the Aral Sea region carry dust towards the south-west and 25% travel westward over the Ustyurt Plateau (Micklin 1988); and Aral dust has been reported as far afield as Belarus and Lithuania to the north-west, Georgia to the west and Afghanistan to the south-east (Létolle and Mainguet 1993).

Orlovsky et al. (2005) give a detailed treatment of the dust storms that occur in Turkmenistan, where the highest frequency (Fig. 5.44) occurs in the Karakum Desert, notably at Repetek (62 days per annum). This is an area where mountains channel strong winds. The plains have the highest incidence of dust storms in the spring months, when the soils dry out and there is a great incidence of energetic cyclones and cold-wave intrusions.

There have been few detailed studies of dust storms elsewhere in Central Asia. North of Turkmenistan, in Kazakhstan, dust-raising occurs in the desert areas between the Aral and Caspian seas (Fig. 5.45). Mineral dust from the Ryn Peski desert, north of the Caspian Sea, has been detected 2000 km distant in countries bordering the Baltic (Hongisto and Sofiev 2004).

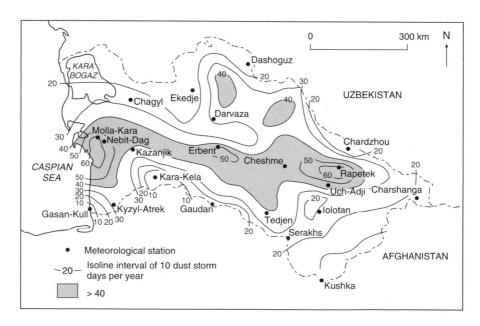

Fig. 5.44. Distribution of dust storms (visibility <1000 m) in Turkmenistan. Modified after Orlovsky et al. (2004, Fig. 2)

Fig. 5.45. A wall of dust approaching the village of Qulandy, north west of the Aral Sea in Kazakhstan, from the desert clay plains of the Ustyurt plateau to the west, 26th July 2004 (from NJM)

5.10 China

Dust storms of yellow dust take on particular importance in China because of their significance for the formation of loess (Derbyshire et al. 1998; Kar and Takeuchi 2004). They also appear to have been a major source of the dust in Late Pleistocene ice layers in Greenland (Svensson et al. 2000) and forested peat bogs from Kalimantan in south-east Asia (Weiss et al. 2002). Moreover, according to Kes and Fedorovich (1976), the Tarim Basin has more dust storms than any other location on Earth, with 100–174 per year. There are stations to the north-west of the 750-mm annual isohyet that have dust storms on more than 30 days in the year (Goudie 1983). These dust events can cover immense areas and transport particles to Japan (where the dust haze is known as *kosa*, literally "yellow sand" in Japanese), Korea (where the dust is called *Huang Sa*), Taiwan and the Pescadores Islands (Yuan et al. 2004) and beyond (Ing 1972; Willis et al. 1980; Betzer et al. 1988; Chung et al. 2003; Fig. 5.46a, b). The dust is highly seasonal in occurrence, with the spring months being the time of greatest activity (Youngsin and Lim 2003; Laurent et al. 2005). The question of the trajectories of long-distance dust derived from China is discussed further in Section 5.12.

Studies of dust loadings (Chen et al. 1999) and fluxes have suggested that there are two main source areas: the Taklamakan and the Badain Jaran

a)

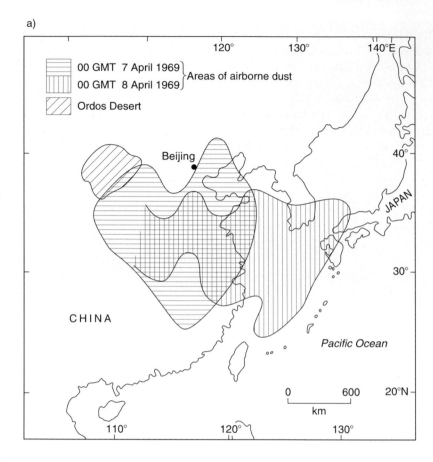

b)

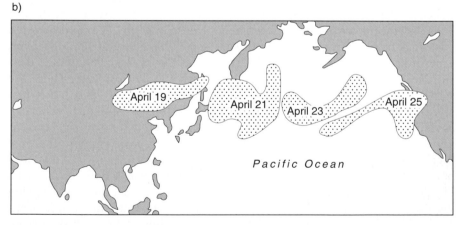

Fig. 5.46. a) The progress and location of a large dust storm over China in April 1968. Modified after Ing, in Willis et al. (1980, Fig. 2). b) The progress of a dust cloud across the Pacific to North America in April 1998. Modified after Husar et al. (2001, Fig. 2)

(Zhang et al. 1998). In all, it has been estimated that about 800 Tg of Chinese dust is injected into the atmosphere annually, which may be as much as half of the global production of dust (Zhang et al. 1997). The prevalence of yellow dust haze in the Tarim basin has been noted by many travellers and, in his *Pulse of Asia,* Huntington (1907, p. 157) reported that: "Dust fell so fast, that even on a still day one was obliged to brush his letter-paper every ten or fifteen minutes to prevent the pen from becoming clogged. Almost every traveler speaks with exasperation or weariness of the persistence with which the haze shrouds the land for weeks at a time".

Figure 5.47 shows two of the best available maps of dust storms in the region. The predominant importance of the Taklamakan (located in the Tarim Basin) is evident, though other important centres occur north of Urumqui in the Junggar Pendi and in the Ordos. Sun (2002a, b) draws attention to the Tengger, Ulan Buh, Hobq and the Mu Us deserts (generally referred to as the 'Gobi Deserts') as sources for the loess deposits of the Chinese Loess Plateau. Indeed, Zhang et al. (2003), Sun (2000) and Xuan et al. (2004) argue that they may be as important as, or even more important than, the Tarim Basin. Shao et al. (2003) concur, suggesting the Gobi to be the strongest dust source in the region, with dust emission rates of up to 5000 $\mu g\ m^{-2}\ s^{-1}$.

The analysis by Shao and Wang (2003) has it that the highest frequency of dust events occurs in the Taklamakan but that most of these events are classified as dust in suspension rather than full-blown dust storms (Fig. 5.48). The Gobi Desert experiences fewer dust events, but they are often severe and extensive. These authors found atmospheric dust concentrations in the Tarim Basin and the Gobi regions to be of a similar order of magnitude, with average maximum values reaching 1 mg m^{-3}.

The TOMS data (Fig. 5.49) confirm the primacy of the Taklamakan/Tarim source. A large area stretching over 75–94° E and 35–42° N has relatively high AI values, which exceed 11 in the centre. The Junggar Pendi shows up as a secondary source, as do some small areas to the east of the Taklamakan towards Beijing. The TOMS mean values are in broad agreement with modelled dust production (Xuan et al. 2000), in that both show an east-to-west increase in dust, with a primary peak in the Tarim and a springtime maximum. However Xuan et al. (2000) suggest a secondary peak over West Mongolia, which is not evident in the TOMS data.

The strength of the Taklamakan as a source is scarcely surprising. It is the largest desert in China, has precipitation that can drop to <10 mm and consists of a closed basin into which mountain-sourced rivers feed sediments. There are extensive marginal fans and areas of dune sand from which dust can be winnowed (Zhu 1984; Wang and Dong 1994; Honda and Shimuzu 1998) and lake sediments associated with the wandering and desiccated lake of Lop Nor. Above all, with an area of 530 000 km^2, the Tarim is one of the Earth's largest closed basins. However, the TOMS data do not indicate that it is a source of similar magnitude to northern Africa. The area with high AI values is both smaller and less intense.

a)

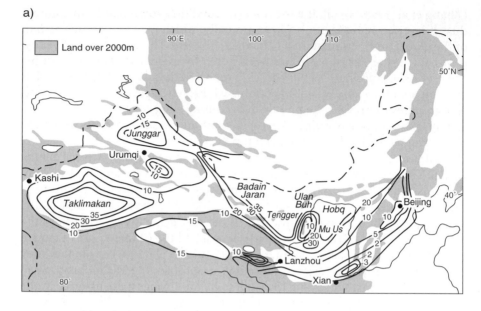

b)

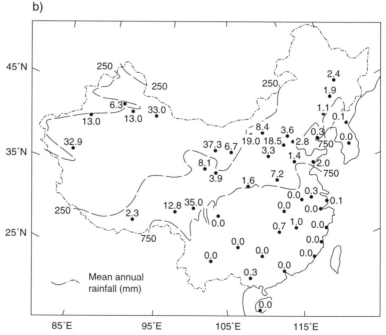

Fig. 5.47. a) The 30-year mean annual number of sand and dust storm days in North China. Source: Derbyshire et al. (1998, fig. 13). b) Distribution of surface-observed dust-storm frequencies in China. Modified after Middleton (1986, Fig. 8.2)

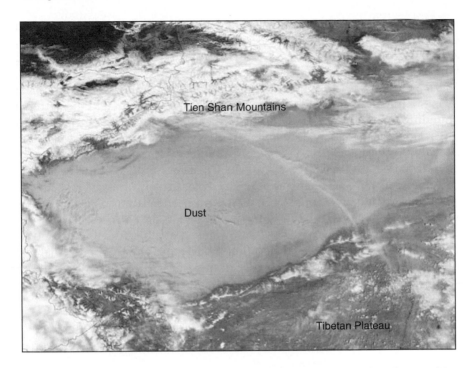

Fig. 5.48. MODIS image of suspended dust filling the Taklamakan desert of North West China, 18 April 2003

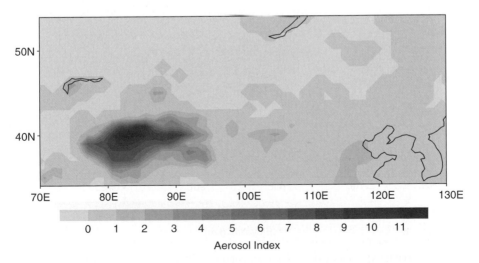

Fig. 5.49. Annual average TOMS AI values for China. Modified after Washington et al. (2003, Fig. 12)

The atmospheric circulation associated with dust of Asian origin has been the subject of numerous studies (see, e.g. Iwasaka et al. 1983; Littmann 1991a; Zaizen et al. 1995; Husar et al. 2001) and is known to be enhanced during the boreal spring (Prospero and Savoie 1989; Jaffe et al. 1997; Talbot et al. 1997, Husar et al. 2001). The circulation over the Taklamakan is highly complex, owing to the influence of the seasonally reversing monsoon and the extreme bounding topography, thereby obstructing any through flow of the prevailing winds. Dust loadings are highest in late winter and spring and are probably associated with cold waves or surges of the north-east monsoon. A local maximum in surface-wind velocity occurs at the southern edge of the Taklamakan, presumably where the cold-air advance is blocked. An additional explanation could be that the dust-laden atmosphere is poorly ventilated, so that dust products remain trapped in the enclosed basin.

Figure 5.50 shows an overlay of TOMS values, potential sand flux (q) and an elevation derived from a digital elevation model at 0.5° resolution. The largest potential sand-flux values in the entire domain (20–50° N, 80–110° E) are in very close proximity to the maximum in AI values (Washington et al. 2003).

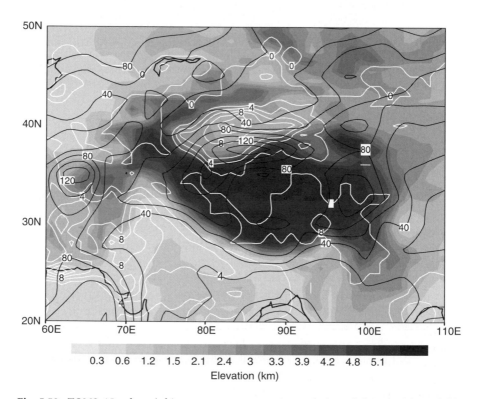

Fig. 5.50. TOMS AI values (white contours, contour interval 2), scaled potential sand flux (black contours, contour interval 20) and elevation in kilometres (grey scale) for China, long-term annual means. Modified after Washington et al. (2003, Fig. 13)

The highest potential sand-flux values are only slightly offset to the south of the AI values and run up against the Tibetan plateau. As in the case of the Bodélé depression in the Sahara, high potential sand-flux values relate to regions of extreme topographic channelling of the winds. In this case, the channelling occurs through one of the largest closed basins in the world.

5.11   Mongolia

The characteristics of dust storms in Mongolia are discussed by Middleton (1991), who found the most frequent activity was in the southern region of the Gobi desert, where Zamiin Uud records an average of 34.4 dust storm days per year. The spring months of April and May are those with the greatest dust storm activity. It is then that average wind speeds reach a maximum and the snow cover is receding. In spring, the Siberian High breaks down and fronts track across the area. The distribution pattern established by Middleton has been confirmed by Natsagdorj et al. (2003), as is the springtime peak of activity.

5.12   Trajectories of Dust Transport from China and Mongolia

Dust from the large expanse of desert across northern China and Mongolia has been found in glacier ice on the northern and western margin of the Tibetan plateau (Wake et al. 1994), but the transport trajectory that has been subject to much greater study is eastward, out towards the Pacific Ocean. Studies of north-east Asian dust outbreaks, occurring particularly in the spring months and that reach Korea and Japan (Fig. 5.46a, b), include those of Chun et al. (2001), Ma et al. (2001) and Mori et al. (2002, 2003), while a study by Osada et al. (2004) examines accumulations in snow banks in Central Japan. This material also commonly reaches the North Pacific Ocean (Duce et al. 1980) and the islands of Hawaii (Shaw 1980) and can travel as far as Alaska (Rahn et al. 1981). Desert dust was found to be the dominant form of aerosol in the middle troposphere, at 5–6 km altitude, north of 23° N, a region of prevailing westerlies, during sampling for the Pacific Atmospheric Chemistry Experiment campaign, in January 1994 (Ikegami et al. 2004).

More south-westerly trajectories also occur and dust from China is often reported from Taiwan (Chen et al. 2004; Wu et al. 2004). This material may also return to the Chinese mainland should synoptic conditions allow, as was observed in Hong Kong in May 1996 (Fang et al. 1999).

Intercontinental transport of mineral dust from some very large dust events in north-east Asia has been traced to North America in recent years (e.g. the April 2001 event; Takemura et al. 2002; Daremova et al. 2005). Material from a major dust storm in northern China in April 1998 was observed on satellite imagery crossing the Pacific over a period of five days (Fig. 5.46b) and being deposited in Canada by large-scale subsidence and orographic effects (McKendry et al. 2001). Dust-raising occurred across an area of about 3.3×10^6 km^2 of northern China for a period of some ten days; and an estimated 4.64×10^8 t of dust was emitted over this period, most of it from the Gobi (In and Park 2003).

A second, even larger trans-Pacific dust transport episode took place in April 2001, following extensive dust-raising over the Taklamakan and Gobi deserts. PM10 concentrations of Asian dust reached 30–40 µg m^{-3} at a large number of rural sites in the United States and contributed to even larger concentrations at some urban locations (Jaffe et al. 2003).

These large trans-Pacific dust events are typical of the spring months, the time of maximum dust storm activity in northern China, and are relatively rare: Jaffe et al. (2003) identified just the two large events mentioned above in 15 years of aerosol observations. However, mineral dust from Asia is transported to North America in smaller quantities all year round. Examination of data from the Interagency Monitoring of Protected Visual Environments (IMPROVE) network by VanCuren and Cahill (2002) found Asian dust at all of the sites in the western United States throughout the year, with a broad maximum between March and October.

Probably much less frequent are events that transport mineral dust from Chinese deserts to Europe. Grousset et al. (2003) found evidence of dust from China in the French Alps, having been transported more than 20 000 km across the North Pacific, North America and the North Atlantic.

5.13   Australia

Like other parts of the Southern Hemisphere, Australia is not an especially dusty continent. However, both at the present and in the past, dust activity has been appreciable and has contributed to sedimentation on- and off-shore (McTainsh 1989; Knight et al. 1995; Kiefert and McTainsh 1996; Goede et al. 1998). Australia is today the largest dust source in the Southern Hemisphere and in the Late Glacial Maximum contributed three times more dust to the South West Pacific than now (Hesse and McTainsh 1999). Notable dust events of the twentieth century included the great 'dust-up' of November 1902, the series of storms that darkened the midday sky in Adelaide in the summer of 1944–1945 and the huge pall of Mallee-derived dust that swept across Melbourne during February 1983 (Lourensz and Abe 1983; Fig. 5.51). On 23 October 2002, Australia's largest reported dust storm caused air pollution across most eastern parts of the country after 12 months of extreme drought combined with above-average maximum temperatures to produce severe soil

Fig. 5.51. Dust storm in Melbourne, Australia, 8 February 1983 (Australian Bureau of Meteorology)

moisture deficits and reduced vegetation cover (McTainsh et al. 2005). The storm measured 2400 km long, up to 400 km across and between 1.5 km and 2.5 km high. The dust load was estimated at between 3.35×10^6 t and 4.85×10^6 t; and it caused air quality problems for the inhabitants of several cities, including Sydney and Brisbane (Chan et al. 2005).

The average distribution of dust storm activity in Australia has been plotted from meteorological data by McTainsh and Pitblado (1987; Fig. 5.52a) and shows six areas of above average activity: Central Australia (A), Central Queensland (B), the Mallee (C), the Eastern and Western Nullarbor plains (both labelled D) and coastal Western Australia (E), with in excess of five storms per year. McTainsh et al. (1989) also subdivide eastern Australia with regard to dust storm seasonality (Fig. 5.53). A northern region (encompassing Queensland and most of New South Wales) experiences dust storms during spring and early summer, whereas in the southern region (southern New South Wales and Victoria) dust storms are prevalent during summer. This relates in part to the rainfall regimes of these two regions, though in both areas the months with most frequent dust storms are also the windiest months as well.

Substantial quantities of dust leave Australia in two main plumes (Fig. 5.52b): one that runs across the Tasman Sea towards New Zealand (McGowan et al. 2000; Marx et al. 2005a) and another that heads westward out into the Indian Ocean (McTainsh 1985). The former plume was more active during the last Glacial Maximum (Thiede 1979) and is an important

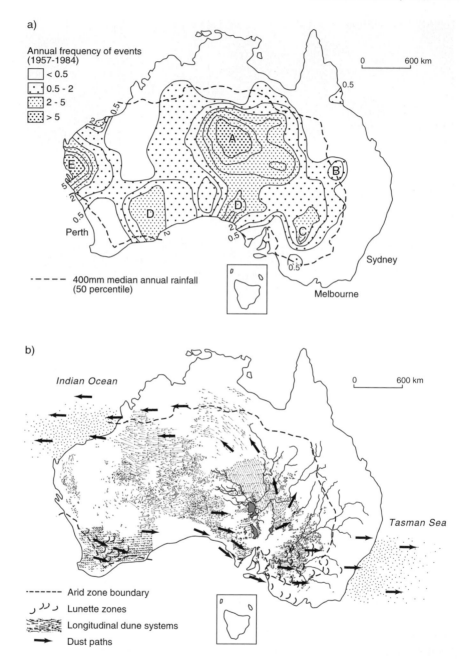

Fig. 5.52. Dust storm activity in Australia. a) Average annual frequency (1957–1984). b) Dust paths into the Tasman Sea and Indian Ocean in relation to aeolian landforms. Modified after McTainsh (1989)

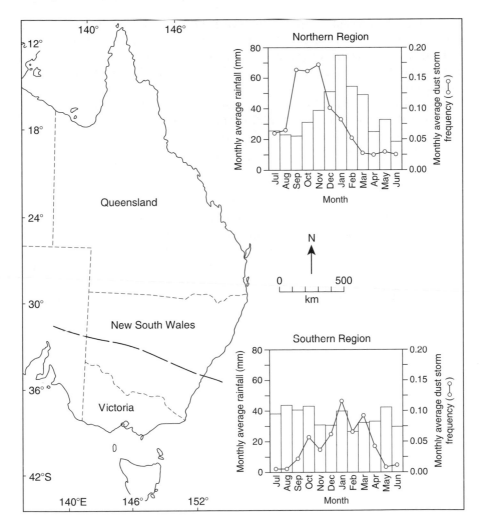

Fig. 5.53. Wind erosion regions of eastern Australia. Modified after McTainsh et al. (1989, Fig. 1)

contributor to Tasman Sea sediments (Hesse 1994), but dust from Australia still reaches New Zealand with some regularity. McGowan et al. (2005) and Marx et al. (2005b) traced New Zealand dust back to the Eyre Peninsula of South Australia and western New South Wales. The Channel Country north of Lake Eyre and the Simpson Desert have probably been major sources of dust in arid phases. Much dust may also be derived from around Lake Eyre and the Murray–Darling plains. Some dust has accumulated on land, contributing to the formation of 'parna', a clay-rich sediment.

The area of greatest dust-storm frequency, as determined from meteorological station data, has been shown to be broadly coincident with the huge

$(1.3 \times 10^6 \text{ km}^2)$ internal drainage basin of Lake Eyre. Indeed, TOMS analysis indicates that this ephemeral playa is the continent's main dust source, the only area where AI values exceed 11. The dustiness of the current playa bed itself can only be inferred from terrestrial data, due to the absence of meteorological stations.

As an area of sediment supply, the Lake Eyre Basin has been compared to that of Lake Chad (McTainsh 1985), with deflation operating on alluvial spreads brought by the southward-flowing Eyre, Diamantina,and Cooper rivers. The long history of deflation is evidenced by wind-blown deposits, typically rich in gypsum and clay, found at a number of sites (Magee and Miller 1998). Indeed, after making an approximate comparison of the sediment yield of dust transport and river systems in the Lake Eyre and Murray–Darling Basins, Knight et al. (1995) summed up the overall significance of dust transport in the evolution of the Australian landscape by asserting that more sediment is lost from the continent in the air than in rivers.

Throughout the months of maximum atmospheric dust-loadings (October to March) the surface-wind speeds reach a maximum over the Simpson and Great Victorian deserts (apart from the west coast of Australia), with a prevailing south easterly to southerly wind. The classic synoptic situation generating deflation in southern regions of Australia is an eastward moving mid-latitude frontal system (Sprigg 1982). Anticyclogenesis may follow the passage of the front, producing marked horizontal wind shear in the easterlies to the south of a heat trough (Sturman and Tapper 1996). Material raised by these systems is occasionally transported as far as New Zealand (Collyer et al.1984; McGowan et al. 2000).

6 Dust Concentrations, Accumulation and Constituents

6.1 Dust Contents of Air

In this chapter we discuss the physical characteristics of dust and deal with such issues as the concentration of dust in the atmosphere, the rates at which it accumulates and the nature of its constituents.

Numerous observations have now been made on the dust contents of air, which help to indicate areas where aeolian material is an important atmospheric component (Fig. 6.1). Duce (1995) provides full details over land and sea, based on measurements in the near-surface boundary layer using high-volume filtration systems. In areas where dust is raised, such as the Thar Desert of north-west India and the Great Plains of the United States, dust concentrations may be in the range from 10^2 µg m^{-3} to 10^5 µg m^{-3}. In the Negev, dust concentrations during dust storms are between 1578 µg m^{-3} and 4204 µg m^{-3} (Offer and Azmon 1994), though Ganor and Foner (2001) record one dust storm in Tel Aviv when a concentration of 23 790 µg m^{-3} was recorded.

At the other end of the scale as, for example, in the North Atlantic between Iceland and Newfoundland, concentrations fall to as low as 0.003 µg m^{-3}. Most oceanic sampling sites are in the range 0.02–1.0 µg m^{-3}. The major exception to this, which once again illustrates the importance of the Sahara as a source of atmospheric dust, is the eastern Atlantic off West Africa where observations indicate dust concentrations from ca. 2.0 µg m^{-3} to ca. 60.0 µg m^{-3}, though in some events airborne dust concentrations off West Africa may reach as high as 13 421 µg m^{-3} (Lepple and Brine 1976), and 13 735 µg m^{-3} (Gillies et al. 1996).

Duce (1995) also recognizes regional differences in dust concentrations over the Pacific. Very low values are found over the equatorial Pacific, central South Pacific and the Southern Ocean, while higher concentrations are evident in the western South Pacific, consistent with moderately high transport from the Australian deserts. The highest concentrations are generally found in the mid- and high-latitude North Pacific, where seasonal transport from the Asian deserts is significant. When Mori et al. (2003) monitored the variation in mass concentration during a long-range *kosa* event emanating from Mongolia in March 2001, they found concentrations dropped by an order of magnitude as the dust was transported across the interior of China (6700 µg m^{-3} at about 500 km from source) to a Japanese island (230 µg m^{-3} at about 2500 km from source). During a dust storm in April 2000, total concentrations

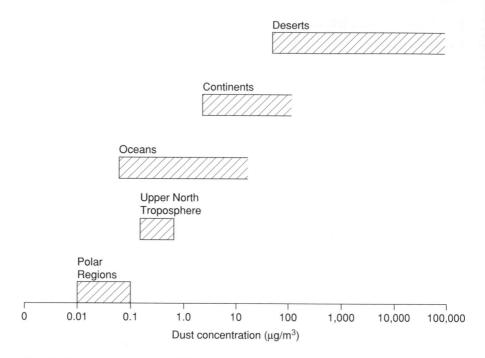

Fig. 6.1. Dust concentrations in different environmental settings. Modified after Schütz (1987, Fig. 3)

in Beijing reached 3906.2 µg m^{-3} (Zhang et al. 2003) while, during a dust event in April, total concentrations at Yulin (600 km east of Beijing) reached 4650 µg m^{-3} (Lasserre et al. 2005).

It is plain that, during dust storm events, levels of particulate matter can often exceed acceptable levels in terms of air quality and health considerations, even at considerable distances from source. For example, during a dust event in Beijing in August 2000 (Xie et al. 2005), average daily PM10 values, that is particles with a diameter <10 µm, reached 720–898 µg m^{-3} which compared with an average daily concentration of 162–190 µg m^{-3} and a Chinese National Ambient Air Quality Standard of 150 µg m^{-3}. In Korea, Chung et al. (2003) found that maximum PM10 values from four dust events between 1997 and 2000 ranged from 254 µg m^{-3} to 996 µg m^{-3}. In the Canary Islands (Querol et al. 2004), daily PM10 values during Saharan dust events can reach up to 1000 µg m^{-3}, which compares with a regional background value of only 19 µg m^{-3}. The European Community Air Quality Directive indicates that daily PM10 values should not exceed 50 µg m^{-3} for more than 7 days year^{-1} or an annual mean of 20 µg m^{-3} (Rodriguez et al. 2001).

The relationship between the mass concentration of dust in the air and visibility is illustrated in Fig. 6.2. Some attempts have been made to relate dust concentrations to the Aerosol Index (AI) determined by TOMS. Alpert and Ganor (2001) suggested the relationship shown in Table 6.1.

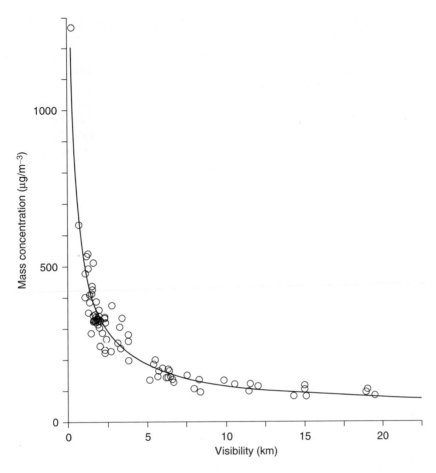

Fig. 6.2. Plot of total mass concentration versus visibility with the corresponding regression curve. Modified after Mohammad and Frangi (1986, Fig. 4)

6.2 Dust Deposition and Accumulation

It is important to distinguish between dust deposition and dust accumulation. As Goossens (2005) has pointed out, deposition refers to the amount of sediment that impacts on a unit surface in a unit time, whereas accumulation

Table 6.1. Relationship between dust concentrations and AI values determined by TOMS

AI value	Daily dust concentration at the surface ($\mu g \ m^{-3}$)
30	4000
25	1900
12	1200

is the amount of sediment that *remains* at a unit surface at the end of a particular time interval.

The information available on the rates of deposition in individual events (Table 6.2) suggests that these can be of a high order. The 1901 dust fall over North Africa, for example, is estimated to have deposited 15×10^7 t and the 1903 dust fall over England is estimated to have deposited ca. 10^7 t of sediment. The range of values lies between 10^5 t and 15×10^8 t. When expressed in terms of sediment deposited per unit area, rates can reach as high as 455 t km^{-2} (North Africa), 300 t km^{-2} (Nebraska), 162 t km^{-2} (Colorado) and 126 t km^{-2} (Caspian) on desert margins. Quantities fall off rapidly towards more humid areas. Data for a major dust storm in the UnitedStates and Canada in 1933, for instance, show deposition rates of 39 t km^{-2} in Kansas, 13.5 t km^{-2} in east Nebraska, but

Table 6.2. Dust deposition rates in individual falls

Location Absolute quantities	Date	Reference	Quantity Tonnes
England	1903	Mill and Lempfert (1904)	10 000 000
England and Wales	1958	Stevenson (1969)	1 000 000
New Zealand	1930	Kidson and Gregory (1930)	200 000
Wisconsin, USA	1918	Winchell and Miller (1918)	>1 000 000
Arctic	1976	Rahn et al. (1977)	500 000
North-west Africa	1974	Lepple and Brine (1976)	400 000
Europe	1901	VDL (1902)	800 000–1 000 000
Europe	1901	Fett (1958)	1 800 000
Sweden	1892	Fett (1958)	500 000
North Africa	1901	Fett (1958)	150 000 000
Poland	1928	Fett (1958)	1 140 000
New Zealand	1928	Fett (1958)	100 000
Kansas, USA	1933	Fett (1958)	131 000
Quantity per unit area			Tonnes km^{-2}
Iowa, USA	1937	Bennett (1938a)	13.2
Michigan, USA	1937	Bennett (1938a)	5.75
New Hampshire, USA	1937	Bennett (1938a)	3.86
Westphalia	1859	Bennett (1938a)	33.1
New Hampshire/ Vermont, USA	1936	Robinson (1936)	3.86
East Nebraska, USA	1933	Weaver and Flory (1937)	13.5
Philadelphia, USA	1934	Watson (1934)	1.35
Ottawa, Canada	1933	Page and Chapman (1934)	1.11

Table 6.2. Dust deposition rates in individual falls—cont'd

Location Absolute quantities	Date	Reference	Quantity Tonnes
Montreal, Canada	1933	Page and Chapman (1934)	0.77
New Hampshire, USA	1933	Page and Chapman (1934)	0.57
Nebraska, USA	1935	Tricart and Cailleux (1969)	300.0
Canary Islands	1973	Logan (1974)	2.0
Canary Islands	1974	Logan (1974)	8.0
Steiermark	1896	Fett (1958)	30.0
Kansas, USA	1933	Fett (1958)	39.0
Caspian Sea	1925	Brouievitch and Goudkov (1954), cited by Tsyganenko (1968)	126.0
South-eastern Australia	1969	Walker and Costin (1971)	19.5–170.0
Europe	1901	Fett (1958)	3.83
North Africa	1901	Fett (1958)	455.0
Poland	1928	Fett (1958)	11.75
New Zealand	1928	Fett (1958)	5.88
England	1903	Mill and Lempfert (1904)	195.0
Egypt	1941	Oliver (1945)	371.0
Sault Ste Marie, Michigan	1937	Martin (1937)	4.5
Marquette, Michigan	1937	Martin (1937)	5.75
Page Co., Iowa	1937	Martin (1937)	30.6
Fort Collins, Colorado	1937	Martin (1937)	162.2
South-eastern France	1846	Free (in Winchell and Miller 1918)	2.2
Salzburg, Austria	1862	Free (in Winchell and Miller 1918)	0.09
Carmiola, Austria	1862	Free (in Winchell and Miller 1918)	5.52
Madison, USA	1918	Free (in Winchell and Miller 1918)	5.21
Idaho, USA	1917	Larsen (1924)	67.2

only 0.58 t km^{-2} in New Hampshire. Nonetheless, moderately high dust falls have been recorded in Europe and Britain (3.83–195.0 t km^{-2}).

When we turn to annual rates of deposition, estimates of rates of dust deposition exist for a number of sites at varying distances from the heart of the Sahara (Fig. 6.3). Others are presented in Table 6.3. As might be expected, there is a tendency for rates to be lowest at large distances from potential sources. Thus the values for Western Europe (e.g. Central France and the Alps) are less

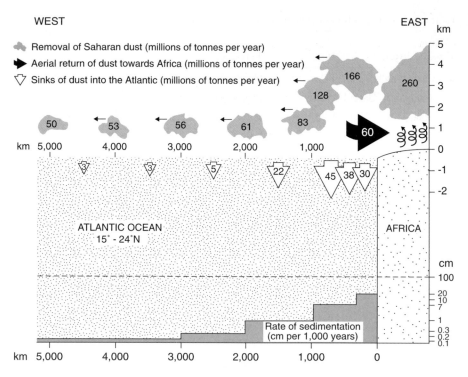

Fig. 6.3. Aeolian sediment budgets for the Sahara. Modified after Schütz et al. (1981, Figs. 8, 9)

than 1 g m^{-2}. Further south, in north-east Spain, a value of 5.1 g m^{-2} is recorded while, in south-east Spain, a value of 23.06 g m^{-2} has been found. Over Sardinia, Corsica, Crete and the south-eastern Mediterranean, most values are between 10 g m^{-2} and 40 g m^{-2} (Fig. 6.4). On the south side of the Sahara, values in areas close to Harmattan source regions have values around 100 g m^{-2} to 200 g m^{-2}, but they decline to low values over the Gulf of Guinea. Deposition rates of Harmattan dust decrease southwards across Ghana (Breuning-Madsen and Awadzi 2005). More general data on rates of deposition are given in Table 6.4.

Most of these data on long-term rates are probably best viewed as rough approximations, since records are in many cases short and because of the difficulties in distinguishing between deposition and accumulation. Other estimates of dust deposition have been gained by modelling (Prospero 1996a), using dust concentration data (Fig. 6.5). The model indicates deposition rates for the Mediterranean of 3–14 g m^{-2} year^{-1}, which are comparable to those obtained from direct measurements. The highest values in the model are for the 10° box at 10–20° N and 20–10° W, with a value of 30.8 g m^{-2}.

Schütz et al. (1981) modelled the annual mass budget of dust transported from the Sahara over the Atlantic in the north-east trade wind zone (see

Table 6.3. Annual dust deposition amounts

Reference	Location	Annual deposition (g m^{-1})
Saharan related		
Drees et al. (1993)	South-west Niger	200.0
McTainsh and Walker (1982)	Northern Nigeria	137.0–181.0
Maley (1982)	South Chad	109.0
Herut and Krom (1996)	Israeli coast	72.0
Herut and Krom (1996)	South-east Mediterranean	36.0
Hernández and Hernández (1997)	South-east Spain	23.06
Breuning-Madsen and Awadzi (2005)	Northern Ghana	20.0
Tiessen et al. (1991)	Northern Ghana	15.0
Bergametti et al. (1989)	Corsica	12.0
Löye-Pilot et al. (1986)	Cosica	12.5
Torres-Padrón et al (2002)	Canary Islands	11.9–30.2
Nihlen and Olsson (1995)	Aegean Sea	11.2–36.5
Pye (1992)	Crete	10.0–100.0
Le-Bolloch et al. (1996)	Southern Sardinia	6.0–13.0
Avila et al. (1996, 1997)	North-east Spain	5.1–5.3
Fiol et al. (2005)	Mallorca	4.5
Measures and Brown (1996)	Gulf of Guinea	3.4–11.5
Bücher and Lucas (1984)	Central France	1.0
Wagenbach and Geis (1989)	Swiss Alps	0.4
De Angelis and Gaudichet (1991)	French Alps	0.2
Non-Saharan related		
Cattle et al. (2002)	Northern New South Wales, Australia	31.4
Singer et al. (2003)	Dead Sea	25.5–60.5
Gill et al. (2000)	Lubbock, Texas	25.0–30.0
Reheis and Kihl (1995)	California and southern Nevada	3.0–30.0

Fig 6.3). A high rate of deposition (up to 20 cm per 1000 years) occurs over the first 2000 km whereas, when most of the mass of dust plume has fallen out (at distances greater than 2000 km), a zone of comparatively low accumulation rates (1–2 cm per 1000 years) occurs. Duce (1995) calculated the mean flux and deposition rates of aerosol minerals over all the oceans (Fig. 6.6, Table 6.5) and, as might be expected, found that the highest flux values, some

Fig. 6.4. Saharan dust on the bonnet of a car in Calvi, Corsica, July 1999 (from ASG)

in excess of 10 000 mg m^{-2} year^{-1} occurred downwind of the major arid regions of Africa, eastern Asia, the Indian sub-continent and Australia. In all he calculated that around 910 Tg year^{-1} of mineral matter was deposited in the oceans, of which just over half was into the North Pacific. More recent estimates yield broadly similar values for the total flux to the global oceans, ranging up to 1814 Tg year^{-1} (Ginoux et al. 2001).

Various estimates have been made of dust deposition rates in China. Ta et al. (2004a), based on 15 years of measurements, came up with average deposition rates of 251.8 kg km^{-2} year^{-1} for the loess region and 365.5 kg km^{-2} year^{-1} for the Gobi desert regions. Deposition rates decline as precipitation amounts increase. Another detailed study was carried out by Zhang et al. (1997). They produced figures of a similar magnitude, with mean regional values ranging from 130 kg km^{-2} year^{-1} to 670 g m^{-2} yr^{-1} (Tables 6.6, 6.7). Rates of deposition of dust in East Asia show a decline with distance from sources in the west of the country. Data from Gao et al. (1997), plotted in Fig. 6.7, show this clearly.

Rates of dust deposition in the Pampas of Argentina were estimated by Ramsperger et al. (1998). They found that dust input was around 400–800 kg ha^{-1} year^{-1}.

However, as will be evident in Chapters 7 and 9, rates of dust deposition show considerable temporal variability at a whole range of time-scales.

Table 6.4. Rates of dust accumulation

Location	Reference	Rate (mm per 1000 years)
Based on measurements made on land		
Iraq	Kukal and Saadallah (1973)	20 100
Caspian Sea	Brouievitch and Goudkov (1954)	659–862[a]
Northern Arabian Sea	Foda et al. (1985)	800
Idaho, USA	Larsen (1924)	500
Illinois, USA	Van Heuklon (1977)	100
Beersheba, Israel	Rim (1952), cited by Yaalon and Ginzbourg (1966)	100
Israel	Yaalon and Ganor (1975)	22–83
Israel	Yaalon and Dan (1974)	20–80
United States	Smith et al. (1970)	65–85
Europe	Free (in Twenhofel 1950)	70
Caspian Sea	Kukal (1971)	46–60[a]
Pyrenees	Bucher and Lucas (1975)	18–23[a]
Kansas, USA	Brown et al. (1968)	6.6–8.6[a]
Adelaide, south-eastern Australia	Tiller et al. (1987)	2.5–5.0
Based on studies of ocean cores and ice caps		
North Atlantic	Prospero and Carlson (1981)	5.0–6.0
East California	Marchand (1970)	2.1–2.6[a]
South-eastern Australia	Walker and Costin (1971)	0.9–1.2[a]
New Zealand	Windom (1969)	0.8
Western tropical Atlantic	Delany et al. (1967)	0.6
North Pacific	Ferguson et al. (1970)	0.1–0.5
Tropical Pacific	Jackson et al. (1971)	0.4
Global	Windom (1969)	0.1–1.0
Global	Judson (1968)	0.25–1.0
Washington State, USA	Windom (1969)	0.21
Greenland	Windom (1969)	0.14
Yukon	Windom (1969)	0.11
Mexico	Windom (1969)	0.01–0.09
Arctic Ocean	Darby et al. (1974)	0.02
Arctic Ocean	Mullen et al. (1972)	0.09
Antarctica	Windom (1969)	0.01

[a] Derived from data expressed as unit mass per unit areas by author, assuming (after Prospero and Carlson 1972) an *in situ* density of 0.65–0.85 g cm^{-2}

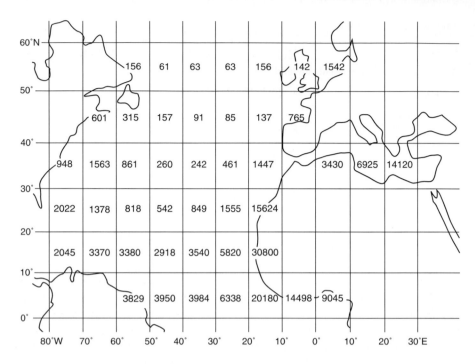

Fig. 6.5. Annual aerosol deposition rates (g m^{-2} × 10^3) over the North Atlantic Ocean. Derived from data in Prospero (1996a, Table 2B)

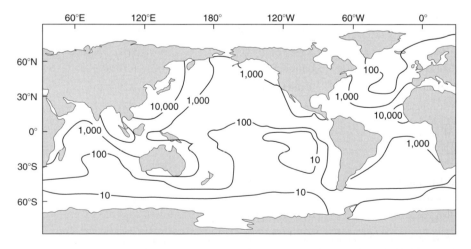

Fig. 6.6. Calculated global fluxes of atmospheric matter (mg m^{-2} year^{-1}) to the oceans. Modified after from Duce (1995, Fig. 3.10)

Table 6.5. Atmospheric mass flux of mineral aerosol to the ocean (from Duce 1995, Table 3.4)

Ocean	Mean flux (g m^{-2} year^{-1})	Deposition (Tg year^{-1})
North Pacific	5.3	480
South Pacific	0.35	39
North Atlantic	4.0	220
South Atlantic	0.47	24
North Indian	7.1	100
South Indian	0.82	44
Global	2.5	910

6.3 Particle Sizes

Some of the earliest determinations of dust deposit grain sizes were made in the United States by Udden (1898). He found that most of his samples were in the size range from 62.5 μm to 15.6 μm. Given that suspension is the prime mode of dust transport, it is to be expected that silt is a dominant component of dust deposits, though clay and sand fractions can also be present. Also, given that coarser particles will drop out of suspension first, dust deposits tend to get finer as one moves away from their source regions. Udden suggested that coarse dust (31–62 μm) might travel around 320 km from its source, that medium dust

Table 6.6. Rates of dust deposition in China. Total atmospheric deposition of mineral aerosol to Chinese deserts. Source: Zhang et al. (1997, Table 1)

Chinese deserts sampled in 1991–1994 (area)	Deposition rate (g m^{-2} year^{-1})	Total deposition (Tg year^{-1})
Taklimakan desert (337 600 km^2)	450 (110–1900)	150.0 (37.0–630.0)
Gurbantunggut desert (48 800 km^2)	130 (37–270)	6.2 (1.8–13.0)
Desert in the Tsaidam Basin (34 900 km^2)	230 (68–480)	7.9 (2.4–17.0)
Kumutage desert (19 500 km^2)	320 (40–1100)	6.2 (0.8–22.0)
Badain Juran desert (44 300 km^2)	310 (99–750)	14 (4.4–33.0)
Ulan Buh desert (9970 km^2)	670 (14–2100)	6.7 (0.1–21.0)
Hobq desert (16 100 km^2)	420 (73–570)	6.7 (1.2–9.2)
Mu Us desert (32 100 km^2)	380 (66–1300)	12.0 (2.1–42.0)
Tengerr desert (42 700 km^2)	290 (15–1200)	12.0 (0.6–52.0)

Table 6.7. Rates of dust deposition in China. The 15-year maximum, minimum and mean annual dust deposition rates in urban cities in Gansu Province, China. Source: Ta et al. (2004a, Table 3)

City	Average (t km^{-2} year^{-1})	Maximum (t km^{-2} year^{-1})	Minimum (t km^{-2} year^{-1})
Desert/Gobi area			
Jiuquan	320.22	539.68	201.57
Zhangye	352.84	542.4	197.96
Jinchang	290.22	621.85	133.14
Wuwei	498.64	688.85	366.57
Average rate	365.48	–	–
Loess area			
Lanzhou	327.02	398.45	249.19
Dingxi	281.63	405.31	181.48
Linxia	256.57	383.93	149.40
Pinliang	256.94	320.24	188.19
Xifeng	207.50	359.93	135.33
Tianshui	180.86	232.46	131.30
Average rate	251.75	–	–

(16–31 μm) might travel around 1600 km from its source and that dust finer than 16 μm might "be largely scattered around the globe".

The wide range of grain sizes that may be present in dust deposits is made evident by particle analysis of dust collected by passive dust samplers in Lubbock, Texas (Gill et al. 2000). Seven samples had a mean clay content (<2 μm) of 23.02%. The average percentage in the size range 2–50 μm was 55.08%. The rest was in the size range 50–2000 μm. The PM10 percentage, which is important from the human health point-of-view, was 47.2%.

The particle size characteristics of Saharan dust are summarized in Table 6.8. It needs to be noted, however, that nearly all the determinations are for dust storms which are not from major source areas and which have travelled outwards into the moister parts of West Africa, to the Atlantic (Stuut et al. 2005), to the Mediterranean or to Europe. It is likely, therefore, that dust storms occurring nearer their source will have coarser grain size characteristics than those listed. Mean modal and median sizes of the travelled dust tend to be fine silt between 5 μm and 30 μm in diameter, though Harmattan dust at Kano (Nigeria) may have a median diameter that reaches 74 μm (McTainsh and Walker 1982), while that from Tanezrouft reaches 72 μm (Coudé-Gaussen 1981). Conversely, samples from southern Ghana, Barbados, Bermuda, the United States and parts of Europe may be less than 5 μm. Although data are sparse, dust storms may transport substantial amounts of clay-sized material (<2 μm).

Fig. 6.7. Rates of dust deposition in China (g m^{-2} year^{-1}). Derived from data in Gao et al. (1997)

Although the modal data presented here are useful, they provide little information about the maximum sizes of grain that can be transported in dust storms derived from the Sahara and other source regions. Schroeder (1985) found aggregated dust particles up to 150 μm in diameter in samples taken on the coastal belt of Sudan, while samples taken over Sal Island

Table 6.8. Particle size characteristics of dust

Reference	Location	Modal, mean or median size (μm)	Clay (%; <2 μm)
Dust from the Sahara			
McTainsh and Walker (1982)	Kano, Nigeria	8.9–74.3 (median)	2.3–32.0
Coudé-Gaussen (1981)	Tanezrouft (central Sahara)	72 (modal)	9.4
Stuut et al. (2005)	Off north-west Africa	8–42 (modal)	–
Khiri et al. (2004)	Morocco	22–37 (median)	–
Coudé-Gaussen (1991)	Maghreb	5–40 (median)	–
Mattson and Nihlén (1996)	Crete	8–30 (modal)	–
Sala et al. (1996)	Spain	4–30 (mean)	–
Criado and Dorta (2003)	Canary Islands	16.9–20.67 (mean)	7.2–9.6
Ratmeyer et al. (1991)	Sal Island	11.9–18.6 (mean)	–
Fiol et al. (2005)	Mallorca	9.3–58.9 (median)	–
Breuning-Madsen and Awadzi (2005)	Ghana	6.8–16.4 (median)	–
Littmann (1991a, b)	West Germany	2.2–16.0 (median)	–
Pye (1992)	Crete	4.0–16.0 (median)	15.0–45.0
Gillies et al. (1996)	Mopti, Italy	16.8 (modal)	–
Ozer et al. (1998)	Genoa, Italy	14.6 (median)	–
Bücher and Lucas (1984)	South-western France	4.0–12.7 (median)	–
Coudé-Gaussen (1991)	South of France	8.0–11.0 (median)	–
Tomadin et al. (1984)	Central Mediterranean	2.0–8.0 (modal)	
Coudé-Gaussen et al. (1988)	Paris Basin (France)	8.0 (modal)	–
Wagenbach and Geis (1989)	Swiss Alps	4.5±1.5 (median)	–
Petit et al. (2005)	Guadeloupe (Caribbean)	4.0 (modal)	–
Talbot et al. (1986)	Barbados	3.2 (median)	–
Arimoto et al. (1997)	Bermuda	2.0–2.3 (mean)	–
Afeti and Resch (2000)	Southern Ghana	1.16 (mean)	–
Perry et al. (1997)	Continental USA	<1.0 (mean)	–
Franzen et al. (1994)	Fennoscandia	2.7 (median)	–
Blanco et al. (2003)	Lecce, Italy	1.7–2.4 (median)	–
Dust from elsewhere			
Chen and Fryrear (2002)	Texas	23.0–35.0 (mean)	–
Osada et al. (2004)	Japan	6.0–21.0 (median)	–
Liu et al. (2004)	China	3.97–93.54 (median)	–

(Cape Verde Islands) off West Africa have yielded individual quartz grains up to 90 μm in diameter and mica flakes up to 350 μm in diameter (Glaccum and Prospero 1980). Prospero et al. (1970) detected individual large particles (>20 μm in diameter) that were carried more than 4000 km from their Saharan source; and Arimoto et al. (1997) recorded particles 43–57 μm in diameter at Bermuda. Saharan dust collected after numerous fallout events over the British Isles has shown that large numbers of so-called 'giant' dust particles (>62.5 μm) are commonly carried more than 3000 km to Northern Europe (Middleton et al. 2001). They have also been found over the Canary Islands and far out into the Pacific, 10 000 km or more from their Chinese source (Betzer et al. 1988). The mechanisms by which such large particles are kept aloft over such large distances are far from clear (see also Section 2.8).

6.4   Dust Chemistry

Because the chemistry of the dust involved in dust storms is important in understanding their possible effects on soils, precipitation chemistry, ocean biogeochemistry and weathering phenomena (see Chapter 2), data are presented in Table 6.9 for 23 samples of dust collected from different parts of the world and derived from different source areas, for which tolerably complete analyses are available. The main component is silica (59.99%), with Al_2O_2 (14.13%), Fe_2O_3 (6.85%), CaO (3.94%), MgO (2.60%), K_2O (2.35%), water and organic matter also being quantitatively important. Mineralogical studies indicate that the great bulk of the silica is made up of quartz.

In Table 6.9, we also present the major element concentrations for Saharan dust as sampled in the southern Sahara/Sahel from the Harmattan source and over Europe. For comparison, figures are given for Chinese and North American (Arizona) dust and for dust storms on a global basis, together with the mean composition of the Earth's surface rocks.

What emerges from these data is that both Harmattan and European dusts are dominated by SiO_2 and Al_2O_3, a characteristic they share with North American and Chinese dusts. The concentrations of these two major elements are similar to those found in world rocks. The dominance of SiO_2 probably reflects the importance of quartz in aeolian dust. Saharan dust also appears to contain appreciable quantities of Fe_2O_3, MgO and CaO, though Harmattan dust is less rich in MgO and CaO than Saharan dust transported northwards to Europe. The $CaCO_3$ content of dust from North African sources has been recognized for its influence in increasing the pH of rainfall in Corsica (Löye-Pilot et al. 1986) and at Erdemli in Turkey (Özsoy and Saydam 2000), while Saharan dust has also been identified as an important source of atmospheric P, mainly insoluble, to the Mediterranean (Mignon and Sandroni 1999).

Table 6.9. Major element analyses of dust. References (1–11): *1* Kano (McTainsh and Walker 1982), *2* Kano (Wilke et al. 1984), *3* Kano (Wilke et al. 1984), *4* Zaria (Wilke et al. 1984), *5* Zaria (Wilke et al. 1984), *6* Italy/central Mediterranean (Tomadin et al. 1984), *7* Italy/central Mediterranean (Tomadin et al. 1984), *8* Pyrenees (Bücher and Lucas 1984), *9* Europe (Bücher 1986), *10* Goudie (1978), *11*= Clarke (1916)

	Harmattan dust (Southern Sahara)					Saharan dust over Europe				Harmattan mean	Europe mean	China mean(10)	Arizona mean(10)	World mean(10)	World rocks(11)
	1	2	3	4	5	6	7	8	9						
SiO_2	66.03	57.19	59.05	57.45	65.04	61.33	51.79	54.83	58.0	60.95	56.49	60.26	57.92	59.9	58.93
Al_2O_3	11.08	12.11	11.32	10.64	9.97	15.52	12.79	16.33	11.0	11.02	13.91	11.40	12.21	14.13	14.98
Fe_2O_3	4.45	5.30	4.63	4.34	3.78	8.06	5.32	6.09	6.0	4.50	6.37	2.91	4.72	6.85	6.1
FeO	–	–	–	–	–	–	–	–	–	–	–	1.37	–	–	–
MgO	0.82	0.81	0.75	0.81	0.62	2.84	3.86	2.90	2.7	0.76	3.08	–	3.01	2.60	3.81
CaO	0.13	3.61	3.01	2.88	1.90	3.47	12.19	10.15	8.6	2.31	8.60	–	2.01	3.94	4.84
Na_2O	0.91	1.46	1.30	2.14	1.12	0.81	1.16	0.98	1.6	1.39	1.14	1.72	1.93	–	–
K_2O	2.04	2.95	2.87	3.26	2.95	3.26	3.26	2.18	1.8	2.81	2.63	2.13	2.63	2.35	2.99
TiO_2	0.73	0.83	0.81	0.82	0.92	0.74	1.01	1.22	1.2	0.82	1.04	0.65	0.74	–	–
P_2O_5	0.17	0.25	0.22	0.18	0.18	0.18	0.42	0.13	–	0.20	0.24	0.19	–	–	–
MnO	0.10	0.08	0.08	0.09	0.08	–	–	0.05	1.6	0.09	–	–	–	–	–
SO_3	–	–	–	–	–	–	–	–	–	–	–	0.20	–	–	–
CO_2	–	4.99	5.47	6.38	4.18	–	–	–	–	5.26	–	–	–	–	–
H_2O	–	9.74	8.94	9.00	7.30	–	–	–	–	8.75	–	0.80	2.14	–	–
LO_i	12.79	–	–	–	–	–	–	–	–	–	–	–	11.64	–	–
Total	99.25	99.32	98.45	97.99	98.04	–	–	–	–	–	–	–	–	–	–

Dust storms can transport appreciable quantities of organic material, because much dead plant debris (leaves, seeds, seed cases, etc.) has a low density and only requires modest threshold velocities for its entrainment. Analysis of Argentinian dust in the Pampas by Ramsperger et al. (1998) revealed that the organic matter content was between 6.7% and 8.3%. Dust samples from Lubbock, Texas, averaged around 10% of organic carbon (Gill et al. 2000) and samples from Tempe, Arizona, averaged 11.6% (Péwé et al. 1981) while, in Australia, Boon et al. (1998) found the organic content averaged 31–34%. Such organic material makes a major contribution to the functioning of dryland ecosystems (Zaady et al. 2001). The organic portion may include large numbers of phytoliths and diatoms (Folger et al. 1967; Abrantes 2003; Breuning-Madsen and Awadzi 2005) and diatomites from desiccated pluvial lakes in the Bodélé depression seem to be major component of the dusts that are generated from the dustiest place on Earth (Giles 2005). Dust may also contain plant waxes (Dahl et al. 2005).

Analyses of samples for constituents, which are not included in Table 6.9, indicate that some other components may locally be important. Bucher and Lucas (1975) found that carbonates amounted to 20–30% in Saharan dust deposited in the Pyrenees; Khiri et al. (2004) found that Moroccan dust had calcite contents that ranged between 23% and 69%; and Alastuey et al (2005) found that Saharan dust deposited on the Canary islands had $CaCO_3$ values of 6–9% and gypsum values of 3.5% to 10.0%. Logan (1974) found that soluble salts in dust landing on the Canaries amounted to 1.2–3.6%; Yaalon and Ginzbourg (1966) reported $CaCO_3$ contents in Negev dust of up to 48% and soluble salts up to 3.1%; Singer et al. (2003) reported that dust collected in the vicinity of the Dead Sea had calcite contents that were between 5.2% and 33.1%, and dolomite contents that were between 11.5% and 14.8%; Modaish (1997) found that the calcium carbonate content of dust in Riyadh, Saudi Arabia, was 31.8%; in Iraqi dust storms Kukal and Saadallah (1973) found that carbonate amounted to 66.1–69.5%; and, around the Caspian Sea, Brouievitch and Goudkov (cited by Tsyganeko 1968) found a calcium carbonate content of 12.3%. Dust samples in Southern Nevada and California contain 8–31% carbonate and 4–19% for soluble salts (excluding gypsum; Reheis and Kihl 1995). Warn and Cox (1951) found that at Lubbock, Texas, carbonate equalled 5–20% and gypsum 5%, while Gill et al. (2000) found they had highly variable calcium carbonate contents of 0.5–15.3%, whereas samples from Tempe, Arizona, (Péwé et al. 1981) had contents that ranged from 1.12% to 3.87%. In northern China, the carbonate content of dust samples ranged from 2.6% to 12.1% (Wang et al. 2005b).

Dust derived from playa (salt lake) surfaces may be rich in soluble salts (Blank et al. 1999); and this is the case with respect to dust blowing off the desiccating floor of the Aral Sea. Australian dusts may contain as much as 50% by weight of salt (Kiefert 1997). This may contribute to groundwater salinity (Acworth and Jankowski 2001). Dust derived from the Owens (dry) Lake includes appreciable amounts of arsenic (Reheis et al. 1999).

6.5   Clay Mineralogy of Dust

The fine fraction of aeolian dust contains various types of clay mineral, which can sometimes give an indication of source regions of dust production. Dusts collected in Texas (Gill et al. 2000) contain three main clay minerals – illite, smectite and kaolinite. Aeolian clay deposits in south-east Australia have kaolinite and illite as their most common clay minerals (Dare-Edwards 1984). In Japan, dust derived from China (Inoue and Naruse 1987) was dominated by kaolinite, illite, vermiculite and montmorillonite. Dust over the Dead Sea contains smectite, kaolinite, illite and palygorskite (Singer et al. 2003).

There are now available a large number of studies of the clay mineralogy of Saharan dust; and Alastuey et al. (2005), for example, found that the three main clay minerals in Saharan dust over the Canary Islands were palygorskite, illite and kaolinite. However, there are major geographical variations in the proportions of different clay minerals derived from different source areas. Caquineau et al. (1998, 2002) detected variations in the clay minerals present in dust collected at Sal Island on the basis of the different source areas from which the dust was derived. Dust originating from the North and West Sahara exhibited the highest amount of illite, whereas kaolinite became predominant when air mass trajectories indicate a Sahelian origin. Kaolinite was dominant in dust originating from the South and Central Sahara, though the amount of illite could not be detected. Such a latitudinal variation in clay mineralogy is consistent with the observations of Chester et al. (1972) along the coast of Western Africa from 25° N to 30° S. Kaolinite concentrations increased towards the equator, whereas illite decreased. Dust samples collected from the Niger and Northern Nigeria also displayed a predominance of kaolinite (Drees et al. 1993; Wilke et al. 1984; McTainsh and Walker 1982).

Along a transect in the Sahara from 19° N to 35° N, Paquet et al. (1984) identified four different groups or sectors. In Northern Algeria, illite and chlorite accounted for around 70–75% of the clay content, kaolinite about 15% and attapulgite 10–15%. Further south, around Beni Abbes and In Salah, attapulgite reached levels of 20–25%. Even further south, around Tamranrassett, Tessalit and In Guessam, illite and chlorite were dominant (60–70%), attapulgite was only 5–10% and kaolinite 25–30%. South of Hoggar and in the Tanezrouft smectites were dominant, followed by kaolinite (20–25%), illite (10–25%), attapulgite (10–15%) and chlorite (5%). They attributed this variability to the nature of the Quaternary sediments and bedrock of the sectors concerned. For example, the sediments of the northernmost zone gain some of their characteristics as a result of the deflation of the inland basins (*Chotts*), while high kaolinite contents may be derived from ancient lateritic weathering profiles.

Sarnthein et al. (1982) also drew a distinction between northern and southern source areas. Dust from the South Sahara and Sahel (south of 20–25° N) is less rich in carbonate but richer in kaolinite and montmorillonite, whereas in the North and Central Sahara, carbonate contents are higher (up to 20–50%) and the dominant clay minerals are illite, chlorite, palygorskite and montmorillonite. Palygorskite has also been recognized as a characteristic mineral of Saharan dust (Coudé-Gaussen and Blanc 1985) reaching Sardinia and the Western Mediterranean (Molinaroli 1996), in dust falling on Skye, western Scotland (Bain and Tait 1977) and in dust reaching the central Mediterranean (Tomadin et al. 1984). At the eastern end of the Mediterranean, kaolinite is a more significant aeolian clay mineral input, its African origin indicated by the northward decreasing abundance in marine sediments (Foucault and Mélières 2000). Dust reaching north-eastern Spain from the Northern Sahara had the following clay minerals: illite, smectite, palygorskite and kaolinite (Avila et al. 1996, 1997), while that reaching Mallorca was dominated by illite, kaolinite and palygorskite (Fiol et al. 2005).

Although the nature of the source region is important for determining the nature of the clay minerals present in dust, gravitational settling effects during transport are of secondary significance (Prospero 1981).

7 Changing Frequencies of Dust Storms

7.1 Introduction

Dust storm frequencies vary at a series of temporal scales. In Chapter 9 we discuss how dust storm activity has changed during the course of the Quaternary era, using evidence from ocean, lake and ice cores and also from the great loess deposits that were laid down by dust storms in the past. This chapter, however, concentrates on the nature and causes of changing dust storm frequencies in recent decades and is largely based on the analysis of climatological data, though attempts have been made to use remote sensing data, such as TOMS, to identify temporal trends (e.g. Barkan et al. 2004).

7.2 The United States Dust Bowl

The Dust Bowl of the 1930s was possibly the most famous case of soil erosion by deflation (Bonnifield 1979; Worster 1979; see Fig. 7.1), though as Malin's (1946) archival studies showed, dust storms were rampant in Kansas in the nineteenth century, long before the sod had been busted by pioneering farmers. In part the Dust Bowl was caused by a series of hot, dry years which depleted the vegetation cover so that, in the words of John Steinbeck (1939, p. 49): "a man didn't get enough crop to plug up an ant's ass". It also made the soils dry enough to be susceptible to wind erosion. The effects of this drought were gravely exacerbated by years of overgrazing and unsatisfactory farming techniques. However, perhaps the prime cause was the rapid expansion of wheat cultivation in the Great Plains. The number of cultivated hectares doubled during the First World War as tractors (for the first time) rolled out on to the plains by the thousands. In Kansas alone, the hectares under cultivation increased from under two million in 1910 to almost five million in 1919. After the war, wheat cultivation continued apace, helped by the developments of the combine harvester and governmental assistance. The farmer, busy sowing wheat and reaping gold, could foresee no end to his land of milk and honey; but the years of favourable climate were not to last and, over large areas, the tough sod which exasperated the earlier homesteaders gave way to friable soils of high wind erosion potential. Drought, acting on

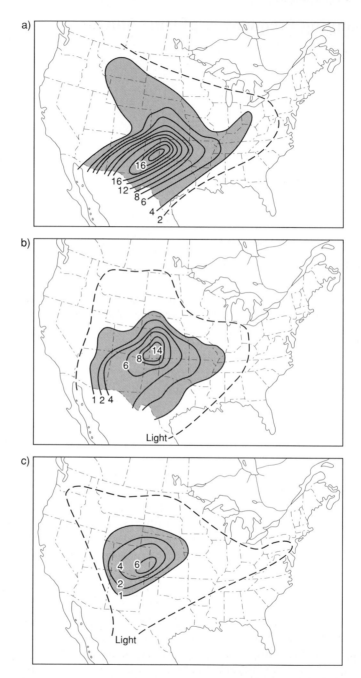

Fig. 7.1. The concentration of dust storms (number of days per month) in the United States in 1936, illustrating the extreme localization over the High Plains of Texas, Colorado, Oklahoma and Kansas: a) March, b) April, c) May. Modified after Goudie (1983)

damaged soils, created the 'black blizzards' (Fig. 7.2) which have been so graphically described by Coffey (1978, pp 79–80):

"There was something fantastic about a dust cloud that covered 1.35 m. square miles, stood three miles high and stretched from Canada to Texas, from Montana to Ohio – a cloud so colossal it obliterated the sky . . . a four-day storm in May 1934 . . . transported some 300 million tons of dirt 1500 miles, darkened New York, Baltimore and Washington for five hours, and dropped dust not only on the President's desk in the White House, but also on the decks of ships some 300 miles out in the Atlantic . . . masses of dust began to billow into huge tumbling clouds ebony black at the base and muddy tan at the top, some so saturated with dust particles that ducks and geese caught in flight, suffocated; some turning the sky so black that chickens, thinking it night, would roost. Oklahoma counted 102 storms in the span of one year; North Dakota reported 300 in eight months."

Woodie Guthrie wrote a song about the Great Dust Storm of 14 April 1935:

"The storm took place at sundown, it lasted through the night,
When we looked out next morning, we saw a terrible sight.
We saw outside our window where wheat fields they had grown,
Was now a rippling ocean of dust the wind had blown.
"It covered up our fences, it covered up our barns,

Fig. 7.2. Dust storm approaching Spearman, Texas, 14 April 1935 (NOAA Photo Library)

It covered up our tractors in this wild and dusty storm.
We loaded our jalopies and piled our families in,
We rattled down that highway to never come back again."

The core of the Dust Bowl area comprised the western third of Kansas, south-east Colorado, the Oklahoma Panhandle, the northern two-thirds of the Texas Panhandle and north-east New Mexico, although most of the Great Plains experienced Dust Bowl conditions at some time during the 1930s. Indeed some of the worst conditions were found as far north as Wyoming, Nebraska and the Dakotas.

The most severe dust storms (the black blizzards) occurred in the Dust Bowl between 1933 and 1938, with activity being at a maximum during the spring of these years. At Amarillo, Texas, at the height of the period, one month had 23 days with at least 10 h of airborne dust and one in five storms had zero visibility (Choun 1936). For comparison, the long-term average for this part of Texas is just six dust storms a year (Changery 1983).

The reasons for this most dramatic of ecological disasters have been widely discussed and blame has largely been laid at the feet of the pioneering farmers and 'sod busters' who ploughed up the plains for cultivation. For although dust storms are frequent in the area during dry years and the 1930s was a period of drought, with high temperatures and low rainfall, the scale and extent of the 1930s events were unprecedented (Fig. 7.3a).

The wave of settlers that arrived in the area from 1914 to 1930, in conjunction with the increasing use of mechanized agriculture, catalysed by high wheat prices, led to exceptionally large-scale wind erosion when drought hit the plains in 1931. In 1937 the US Soil Conservation Service estimated that 43% of a 6.5×10^6 ha area in the heart of the Dust Bowl had been seriously damaged by wind erosion.

An approximate 22-year drought cycle has been identified in the Western USA (Mitchell et al. 1979). Major droughts have occurred in the Great Plains in the 1890s, 1910s, 1930s, 1950s and 1970s; and these droughts are normally periods of exaggerated dust-storm activity (Fig. 7.3b). Soil loss in the 1970s was on a scale comparable to that of the 1930s (Lockeretz 1978). This is significant for, as Gillette and Hanson (1989) observe, the early 1970s was not a period when cumulative departures from normal rainfall were as marked as they had been in the 1930s and 1950s in the Great Plains. Thus other factors such as the occurrence of strong, erosive wind storms may be as important a causative factor as simple lack of rainfall.

Land management techniques rather than climate are probably important in determining the variability of dust storm occurrence (Lee et al. 1993); and Todhunter and Cihacek (1999) have documented a decline in dust storm occurrence in North Dakota (Fig. 7.4), which they attribute to the adoption of practices such as planting of shelterbelts, conservation tillage, crop residue management and land retirement programmes. The decline in dust storm occurrence for the southern High Plains (Stout and Lee 2003) since the 1940s is shown in Fig. 7.5.

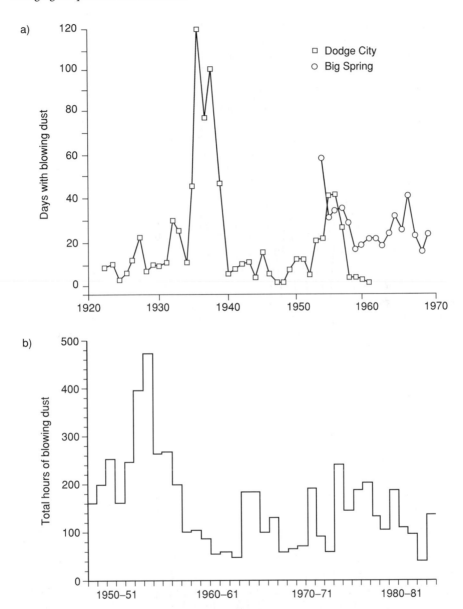

Fig. 7.3. a) Frequency of dust-storm days at Dodge City, Kansas (1922–1961) and at Big Spring, Texas (1953–1970). Modified after Gillette and Hanson (1989). b) Yearly total hours of blowing dust for Lubbock, Texas (summing August through July of the following year). Modified after Wigner and Peters (1987)

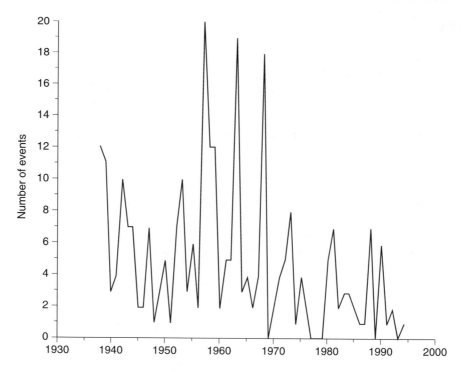

Fig. 7.4. Number of dust events at Fargo, North Dakota. Modified after Todhunter and Cihacek (1999)

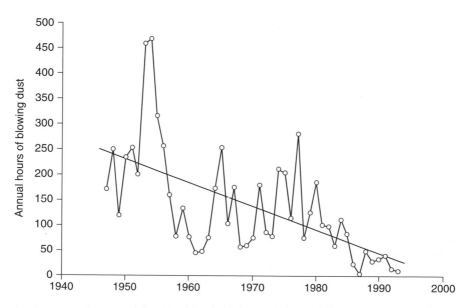

Fig. 7.5. Annual hours of blowing dust from 1947 to 1993 as reported by surface weather observers at Lubbock, Texas. Modified after Stout and Lee (2003, Fig. 2)

7.3   Mexico

A study by Jauregui (1960) of dust storms at Tacubaya in Mexico City over 1923–1958 showed a marked inter-annual variability in frequency, but with no trend over the period. He pointed out in another paper (Jauregui 1969) that dust-storm intensities are generally lower at Tacubaya than at the airport on the northeast edge of the urban area; but as Jauregui's data (1960) for Tacubaya have no apparent visibility limit, it is not possible to compare his frequencies with those taken from the airport. Nevertheless, he found that there was no trend in dust storm frequency from 1923 to 1958, whereas a marked downward trend was apparent for all visibility classes used in the data from the airport over the period 1952–1983 (Fig. 7.6). This decrease in frequency can be explained with reference to three major factors: rainfall, urban development and lake bed management.

A gradual increase in rainfall over the data period is shown by precipitation figures from Tacubaya. A similar steady increase has been noted at San Juan Aragon, about 7 km north-north-west of the airport and on the same side of the urban area, annual rainfall totals here being of the order of 200 mm less than those at Tacubaya (Jauregui and Klaus 1982). Jauregui (1960) found little correlation between one or two year's rainfall and the following year's dust storm frequency, so that a year or two of low rainfall did not necessarily result in a high number of dust storms the following year. For the

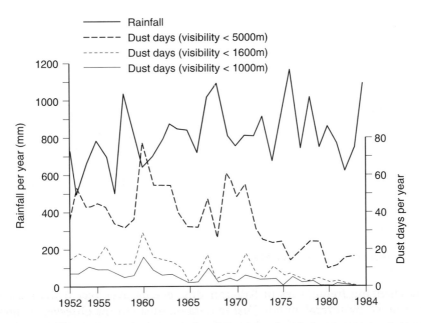

Fig. 7.6. Annual rainfall totals and dust-day frequencies for Mexico International Airport (1952–1984). Modified after Goudie and Middleton (1992, Fig. 2)

data period analysed at the airport, correlations with antecedent annual rainfall were also poor. Comparing annual dust-event frequency (visibility <5000 m) with the previous year's rainfall produced a Spearman's rank correlation coefficient of −0.14 (56% probability of a relationship) and a correlation with the average annual rainfall over the preceding three years was also poor: $r = -0.22$ with 23% probability. Nevertheless, a gradual increase in rainfall can reasonably be expected to have led to a concomitant decline in duststorm frequency over the same period when considered with the other factors outlined below.

The growth and spread of the urban area around the airport may have had some impact on dust storm activity. Although perhaps in the short term urban sprawl may act to further destabilize dust sources during phases of construction, the encroachment of the urban area around the airport that has occurred since the 1950s will have acted to protect susceptible soil surfaces after the initial stages (Jauregui 1989).

The identification of the dried bed of Lake Texcoco as a major source of dust storms resulted in the initiation of a project in 1972 that aimed to stabilize this area. Lake Texcoco, at low levels during dry years, has been a source of dust storms over Mexico City for at least 100 years (Jauregui 1960), but drainage for industry, agriculture and human use resulted in its complete desiccation in the early 1950s. During this decade the lake, situated immediately north-east of the present airport, became a major source of dust storms over both the airport itself and the city. The project has planted pastures of zacate salado (*Distichlis spicata*) irrigated with recycled urban waste water and has constructed a number of reservoirs on the former lake bed. In 1971, the area accounted for 40% of Mexico City's dust storms (SARH 1985); that percentage was reduced to zero by 1984. It is interesting to note that the average annual number of dust storm days occurring at Mexico City shown by Goudie and Middleton (1992) is very much less than that quoted by Jauregui (1973) and subsequently referred to in a number of more general texts on dust storms (e.g. Goudie 1978; Coudé-Gaussen 1984). The value used here is 4.5 days year^{-1}. Jauregui's (1973) value of 68 is quoted from a previous paper (Jauregui 1960) which, as noted above, uses no visibility limit in the dust storm data. Further, Jauregui (1973) uses the value 68 as the average number of dust storms, whereas in the original 1960 paper, he shows an average 68 dust storm days.

7.4   Saharan Dust Events

The changing frequencies of Saharan dust events over recent decades has been noted by several authors, using data on dust storms observed at meteorological stations, satellite observations (e.g. Barkan et al. 2004), data on atmospheric dust concentrations and dust fall deposition rates monitored at distance from source areas (e.g. Chiapello et al. 2005). Increases in dust storm frequency concurrent with drought periods have been noted in the Sahelian zone since the mid-1960s by Middleton (1985a) and by Goudie and Middleton (1992), using data from Mauritania, Senegal, Nigeria and Sudan. N'Tchayi et al. (1997) have also shown an increase in both the frequency of occurrence and annual duration of dust conditions since the late 1950s, particularly for stations in the Sahel; and (N'Tchayi et al. 1994) have demonstrated that as rainfall has diminished the frequency of dust haze has increased (Fig. 7.7a). These trends have been reflected in rising concentrations of Saharan dust monitored at Barbados between 1965 and 1992 (Zhu et al. 1997) and subsequently (Chiapello et al. 2005). The Barbados dust concentrations are inversely related to the previous year's rainfall in Sahelian Africa (Prospero 1996b), but winter transport to that island is related to the North Atlantic Oscillation (NAO) as well (Chiapello et al. 2005).

Atmospheric dust loadings are a function of several climatic parameters that operate on the decadal scale, including drought as mentioned above but also the deflational power of the wind. In the Sahel there is some evidence that this increased between 1970 and 1984 (Clark et al. 2004). Another climatic forcing factor that has attracted recent attention, as mentioned above, is the NAO. Moulin et al. (1997) showed that, between 1982 and 1996, there was a clear similarity in trends between atmospheric optical depth, dust concentrations and the NAO index (Fig. 7.7b). Variations in dust event frequencies could be an indicator of climatic change and this aspect has attracted the attention of several studies in recent years. Observations in north-western Italy have shown an increasing trend of Saharan dust events since 1975 (Rogora et al. 2004). Data from the Mediterranean coast of Spain, south of Alicante, over the period 1949–1994 also showed a marked increase in the number of dust-rain days since the 1970s (Sala et al. 1996). The long-term average there was approximately two dust-rain days per year, but from 1985 to 1994 the annual total averaged 6.5 dust-rain days, with 9.0 dust-rain days per year recorded for the period 1989–1994. In Mallorca, there has also been an increase in dust rains over the period from 1981 to 2003, except for a decrease between 1991 and 1996 (Fiol et al. 2005). Several other authors have remarked upon the peak in Saharan dust falls over Europe in the late 1980s. Dessens and Van Dinh (1990) noted a marked increase in the frequency of Saharan dust outbreaks depositing at the Midi-Pyrenees Aerology Observatory in Lannemezan, France, over the period 1983–1989. Similarly, a significant increase in the quantities of Saharan dust falling over the French Alps since the early 1970s (with very high inputs occurring after 1980; De Angelis and Gaudichet 1991) was detected from an ice core that yielded dust deposition data over a 30-year period (1955–1985). Nonetheless, the 1980s increase was noted in the British Isles (Burt 1991b), which derives Saharan dust both from trans-Mediterranean trajectories and from transport across the Bay of Biscay. Table 5.7 shows the Saharan dust falls over British Isles in the twentieth century documented in the literature, which also affirms the

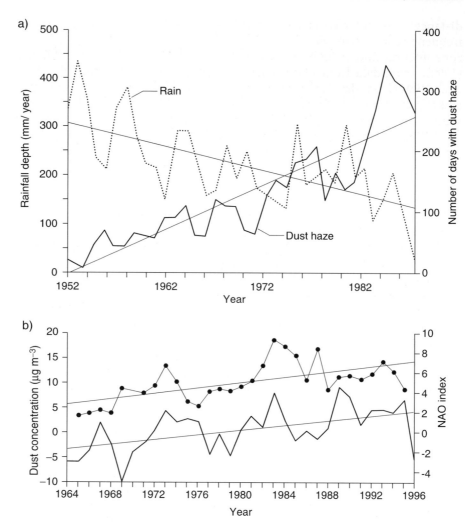

Fig. 7.7. a) The relationship between annual dust haze days and annual rainfall for the Sahelian station of Gao (16° N), between 1952 and 1987. Modified after N'Tchayi et al. (1994, Fig. 6). b) Comparison of the North Atlantic Oscillation (*NAO*) index (*bold continuous line*) with desert dust concentrations at Barbados in the West Indies, between 1964 and 1996. Modified after Moulin et al. (1997, Fig. 4)

importance of the 1980s and early 1990s, although the increase discernible here may also reflect to some extent a keener awareness and interest in such phenomena.

Additional evidence for recent increasing Saharan dust-raising activity comes from the eastern Mediterranean, where Yaalon and Ganor (1979) estimated that some 25×10^6 t of Saharan dust reached the east Mediterranean Basin each year, most settling into the Mediterranean Sea. This figure has subsequently been revised upward, to 70×10^6 tyear^{-1} (Ganor and Mamane 1982) and more recently to 100×10^6 tyear^{-1} (Ganor and Foner 1996). The increase reflects the steady rise in frequency of Saharan dust episodes over Tel Aviv (Israel) from ten per year in 1958 to 19 per year in 1991 (Ganor 1994).

However, it seems that the frequency of 'red rain' events on the Spanish Mediterranean coast has declined in the second half of the 1990s (Avila et al. 1997; Avila and Peñuelas 1999). Eleven years of deposition records (1984–1994) at Corsica showed that annual rates peaked in the late 1980s and declined in the first half of the following decade (Löye-Pilot and Martin 1996). This study also noted the high year-to-year variability, with the annual input of Saharan dust at Corsica varying between 4.0 g m^{-2} and 26.2 g m^{-2} over the study period. Contrary to this evidence of increasing frequency of dust outbreaks across the Western Mediterranean, however, Conte et al. (1996) show a decline in the frequency of strong Siroccos over the period 1951–1990 at Trapani in Sicily. This is probably due to an increase in anticyclonic activity in the western and central parts of the Mediterranean Basin, which tends to counteract the occurrence of frontal disturbances which generate the strong, dust-laden southerly winds from the Sahara.

Data from Mauritania were supplied by the Service Météorologique, Nouakchott. The variation in frequency of annual dust storm days and annual rainfall totals for Nouakchott is shown in Fig. 7.8. The increase in dust storm days after 1968 is dramatic. Low rainfall totals of 48.1 mm in 1970 and 17.9 mm in 1971 represented just 32% and 12% respectively, of the 1949–1967 average and can be seen as the main onset of the drought. The number of dust

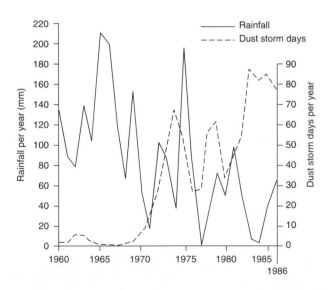

Fig. 7.8. Annual frequency of dust-storm days and annual rainfall for Nouakchott, Mauritania, 1960–1986. Modified after Goudie and Middleton (1992, Fig. 5)

storm days increased markedly from six in 1970 to 65 in 1974 before a reasonably high annual rainfall of 190.6 mm in 1975; dust storm activity declined to 25 days in 1976 and 27 days in 1977. In 1977, however, the rainy season brought just 2.7 mm of precipitation, making it the driest year since records began in 1931; and dust storm activity rose to 55 days and 61 days in 1978 and 1979, respectively. The total dropped to 33 dust storm days in 1980 after a relatively heavy rainfall in 1979, but rose to an unprecedented 85 days in 1983 and remained at around 80 days year^{-1} until 1986.

At Nouakchott, further investigation of the dust storm/rainfall relationship has been undertaken using linear-correlation techniques (Middleton 1986c). The linear-correlation coefficient between dust-storm frequency and the previous year's rainfall (note that the dust-storm season at Nouakchott is primarily in the first six months of the year, before the onset of the rainy season) is −0.53 (28% explanation). The relationship between dust storms and antecedent rainfall is stronger, however, when annual dust-storm days are compared to the average annual rainfall over the previous three years (linear-correlation coefficient = −0.75, with 56% explanation). This fairly strong relationship is similar to that found by Bertrand et al. (1979) for Agadez and Bilma in Niger.

The early 1970s peak in dust-raising at Nouakchott is discernable in data from the north-western margins of the Sahara at Ouarzazate on the headwaters of the Dra in Morocco (Fig. 7.9). Although rainfall was below average in 1971, 1973, 1974 and 1975, other periods of below-average rainfall in the 1960s produced less remarkable peaks in dust storm activity, suggesting a weaker relationship between rainfall and dust-raising in this area, the reasons for which deserve further investigation.

Data for the Sudan refer to stations across the Sudanese Sahel (see Hulme 1985; Middleton 1985a, b). The annual dust-storm frequency and annual rainfall totals for El Fasher, El Obeid and Khartoum (Fig. 7.10) show a marked rise in dust-storm activity dating from the late 1960s/early 1970s. Particularly low rainfall in 1972 and 1973 at El Fasher, for example, was followed by a distinct rise in dust-storm frequency, peaking in 1974, falling in 1975 and 1976 after high rainfall in 1974, but remaining at increasing levels after that year as annual rainfall remained for the most part below 200 mm. The zero dust-storm reading for 1979 followed the wettest year in the central Sudan (1978) in the previous 20–25 years (Trilsbach and Hulme 1984), although particularly high rainfall was not evident at El Fasher itself.

Ozer (personal communication) has analysed long-term visibility and wind data for the Sahara, dating back to the late 1940s. By incorporating such data into D'Almeida's (1986) model of dust emissions, Ozer has come up with some remarkable data. He has calculated that, from the late 1940s to the 1960s, there was a yearly dust production of around 126×10^6 t, which rose to 317×10^6 t during the 1970s and has been at around 1275×10^6 t since the 1980s. In the bad drought year of 1984, dust production reached a startling 3760×10^6 t.

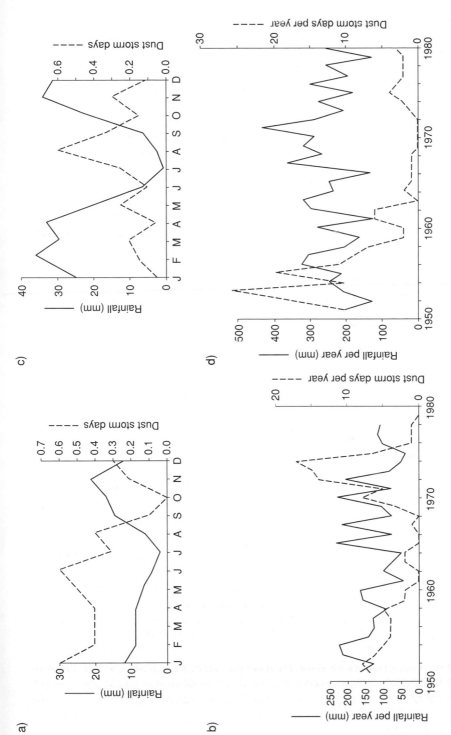

Fig. 7.9. Annual rainfall totals and dust storm frequencies for Ouarzazate and Marrakech, Morocco (1951–1980). a) Mean monthly dust storm frequencies and rainfall for Ouarzazate, b) 1951–1980 annual totals for Ouarzazate. c) Mean monthly dust storm frequencies and rainfall for Marrakech, d) 1951–1980 annual totals for Marrakech

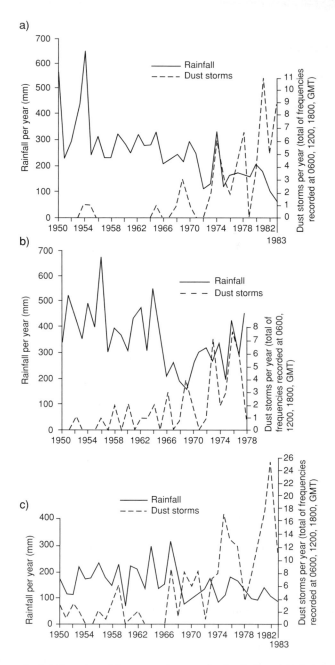

Fig. 7.10. Annual rainfall totals and dust storm frequencies for Sudan: a) El Fasher (1950–1983), b) El Obeid (1950–1978), c) Khartoum (1950–1983). Modified after Goudie and Middleton (1992, Fig. 6)

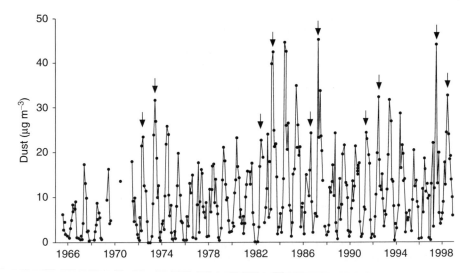

Fig. 7.11. Monthly mean dust concentrations on Barbados (1965–1998; μg m^{-3}). *Arrows* indicate the years when a major ENSO event occurred. Modified after Prospero and Lamb (2003, Fig. 1)

The impact of increased dust loadings over the Sahara in recent decades is also evident in the record of dust transported to Barbados in the Caribbean (Prospero and Lamb 2003). Records have been kept since 1965 and demonstrate a strong correlation with rainfall deficits in West Africa and also with major El Niño events (Fig. 7.11).

7.5   Russia and its Neighbours

Two examples of changing dust storm frequencies can be given for the former USSR; and both demonstrate the important role played by human activities. The first of these relates to the effects of vegetation removal and ploughing in the 1950s as part of the Virgin Lands Scheme when about $40{\times}10^6$ ha of steppe pastures were converted to cropland in eastern Russia, western Siberia and Kazakhstan. As Table 7.1 shows, dust storm frequency in the Omsk region went up on average by 2.5-fold when comparing the data for the period 1936–1950 with those for 1951–1962.

The 1950s also saw a concerted effort to increase the area of irrigated cropland in what was then Soviet Central Asia. In some of these areas with enhanced vegetation cover, the impact on wind erosion has resulted in a declining trend in dust storm occurrence, as shown for two meteorological stations in Uzbekistan in Table 7.2. Simultaneously, the offsite impact of the increase in irrigated cropland has meant a gradual desiccation of the Aral Sea, which has generated great concern about the increasing deflation of dust

Table 7.1. The effects of the Virgin Lands Scheme on the frequency of dust storm days in the Omsk region of the former USSR. After Sapozhnikova (1973)

Station	Mean annual number of dust-storm days		
	1936–1950	1951–1962	Increase
Omsk steppe	7.0	16.0	×2.3
Isil'-Kul'	8.0	15.0	×1.9
Pokrov-Irtyshsk	4.0	22.0	×5.5
Poltavka	9.0	12.0	×1.3
Cherlak	6.0	19.0	×3.2
Mean value	6.8	16.8	×2.5

Table 7.2. Changes in the annual frequency of dust storms at two stations in Uzbekistan due to the expansion of irrigation. After Molosnova et al. (1987)

Station	1950–1959	1960–1969	1970–1979
Khiva	11.9	10.7	5.8
Takhiatash	99.8	34.4	24.8

from the exposed sea bed (Létolle and Mainguet 1993; Middleton 2002). These lacustrine sediments, which are both highly saline and toxic, have become a significant new source of wind-blown material in the region (Fig. 7.12). Marked increases in the annual frequency of dust storms were recorded at several stations in the Priaralye region of Uzbekistan in the following decades, although the rising frequency was not constant at all stations (Table 7.3). Major storms first became visible on satellite imagery in 1975 and have since increased in frequency and duration (Micklin 1988; UNEP 1992). In a summary of estimates of the amount of material deflated annually, Glazovsky (1995) suggested a range of $40–150 \times 10^6$ t as reasonable for the early 1990s.

Orlovsky et al. (2005) studied the annual variation in dust-raising events in Turkmenistan between 1936 and 1995 (Fig. 7.13). Plainly there is a great deal of variability from year to year and decade to decade, but no overall regional picture is evident, except a sharp decrease in frequency after 1980–1985. These authors suggest that a similar fall in dust-raising activity was also recorded for other areas of Central Asia, including the Aral Sea region. Changes in the frequency of latitudinal circulation, irrigation and grazing over the period have all played a role.

Nonetheless, salts from the Aral Sea's exposed bed were recorded as being deposited at annual rates of 0.3 t ha^{-1} in several zones up to 75 km south of the coastline in 1985 (UNEP 1992). A similar deposition monitoring survey in the year 2000 (Wiggs et al. 2003) reported rates in the dustiest period

Fig. 7.12. Dust being raised by north-easterly winds from the desiccated former sea bed of the Aral Sea, 18 April 2003 (MODIS)

of about 0.25 t ha^{-1} month^{-1}. Sodium chloride and sodium sulphate are particularly toxic to plants; and there is a widespread belief that this aeolian deposition of salts is adversely affecting both croplands and natural ecosystems in the Priaralye. Babaev (1996), for example, reports a steady decline in yields of pasture on the Ustyurt Plateau since the 1970s; and Glazovsky (1995) suggests that aeolian salt deposition may at least partly explain decreasing production of silkworm cocoons in the Uzbek autonomous republic of Karakalpakstan.

There is also a possible link between enhanced levels of atmospheric dust and poor human health in areas bordering the Aral Sea. Wiggs et al. (2003)

Table 7.3. Changes in the annual frequency of dust storms at stations in Priaralye, Uzbekistan due to desiccation of the Aral Sea. After Molosnova et al. (1987). *n/a* Data not available

Station	1941–1949	1950–1959	1960–1969	1970–1979
Chabankazgan	n/a	18.1	31.1	44.5
Chimbai	10.6	11.8	13.6	15.0
Muynak	42.2	57.1	18.4	67.6
Zaslyk	n/a	1.5	3.5	12.7

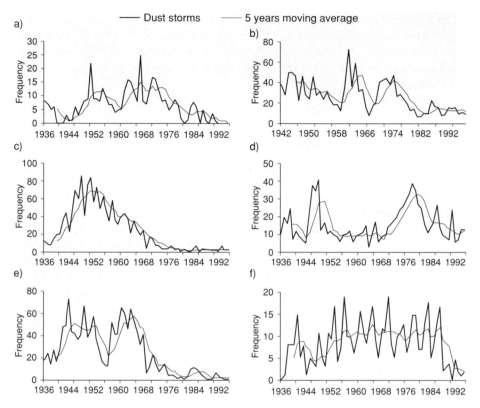

Fig. 7.13. Annual dust storm frequencies (days) for: a) Gasan-Kuli, b) Darvaza, c) Tedjen, d) Dashoguz, e) Kazanjik, f) Kara-Kala. Modified after Orlovsky et al. (2005, Fig. 4)

reported that, in Turkmenistan, respiratory diseases are a major cause of illness and death amongst all age groups and that 50% of all reported illnesses in children are respiratory in nature. Further, there is anecdotal evidence of the emergence of interstitial lung disease amongst children in the region of Kazakhstan that borders the Aral Sea. However, the few studies that have set out to examine these possible links between atmospheric dust and respiratory problems have concluded that dust is just one of several factors that adversely affect respiratory health in the Aral Sea region.

7.6   Pakistan

The frequency of dust storms between 1961 and 2000 has been studied for Pakistan by Hussain et al. (2005). Overall, dust storm frequencies declined in the period 1991–2000 compared to the mean for the whole period, with a

22% decrease in Punjab, 34% in the North West Frontier Province, 45% in Sind and 48% in Baluchistan. This may be attributable in part to the spread of cultivated and irrigated land, though in the late 1990s dust storm incidence appears to have increased again in response to intense drought conditions.

7.7   China and Mongolia

Lake sediments in Korea have been used to construct a mid- to late Holocene history of dust events in China, which has shown that dust storm activity was greatest at times of aridity and strong winter monsoon strength (Lim et al. 2005). In addition, some of the earliest written records of dust storm activity anywhere in the world are recorded in the ancient Chinese literature. They refer to dust falls in northern China, which are variously know as 'dust rain', 'dust fog' or 'yellow fog', usually occurring in the spring months. The earliest known record of 'dust rain' was in 1150 BC and is found in an historical book: *Zhu Shu Nian* ('Chronicles Recorded on Bamboo Slips', quoted by Liu et al. 1981). Written records of dust events in Korea extend back to AD 174 (Chun 2003).

Zhang (1985) used 1156 historical records to show the decadal frequency of dust-rain years in China for the period since 300 AD (Fig. 7.14). The periods of frequent occurrence are 1060–1090 AD, 1160–1270 AD, 1470–1560 AD, 1610–1700 AD and 1820–1890 AD (Liu et al. 2004). Comparison of the frequency of dust-rain years with a winter-temperature index for the period 1470–1969 shows that they are in opposite phase. Although the data set is extensive, it is not evident how homogeneous it is through time; but the period of enhanced dust-raising activity in the nineteenth century has also

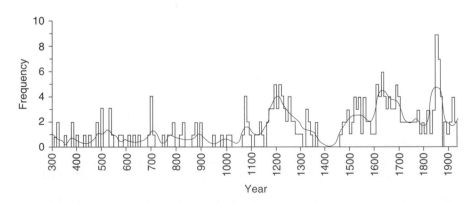

Fig. 7.14. The frequency curve of dust fall since 300 AD in China. Modified after Zhang (1985)

been highlighted from deposits in the Far East Rongbuk ice core near Mount
Everest. Shichang et al. (2001) report markedly more intense dusty periods
during the 1830s, 1840s and 1880s, although the source of the material
found in the core is still open to debate. Dust may have been transported
from northern China, but may equally have been raised locally, from the
Qinghai–Tibetan Plateau, or from south-west Asia. Indeed, the diary of
Tonghe Weng provides data on dust events in Beijing during the late nineteenth century and indicates that their frequency was not markedly different
from today (Fei et al. 2005).

The changing frequency of dust storms during the recent period of meteorological observations has been discussed by a number of authors (see, for
example, Wang et al. 2004b). In Mongolia, Natsagdorj et al. (2003) analysed
data for the period 1960–1999 and identified an increasing trend from the
1960s to the 1980s, with an approximately three-fold increase over that
period, followed by a downward trend in the 1990s (Fig. 7.15a). They believe

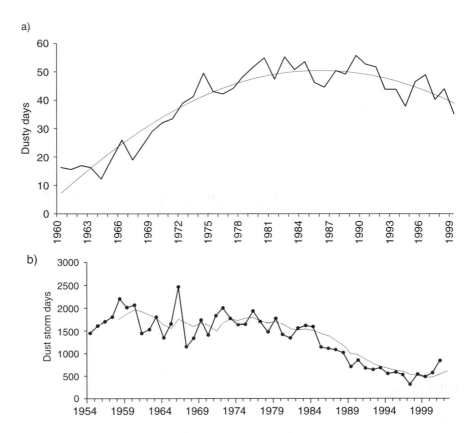

Fig. 7.15. Annual frequency of dusty days in: a) Mongolia, 1960–1999 (modified after Natsagdorj
et al. 2003, Fig. 11), b) China, since 1954 for 681 stations (modified after Wang et al. 2004b, Fig. 6),

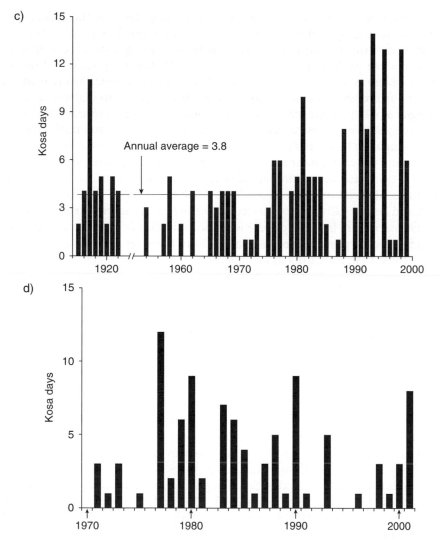

Fig. 7.15. (*continued*) c) Seoul, 1915–1999 (modified after Chun et al. 2001, Fig. 4), although data are missing for 1923–1953, d) Yamanashi Prefecture, Japan, 1970–2001 (modified after Kyotani et al. 2005, Fig. 7)

that human activities accounted for the first of these two phases, but that an increase of precipitation may have caused the reversal in trend during the latter phase.

Zhou and Zhang (2003) and Wang et al. (2005a) analysed the frequency of severe dust storms for the period since 1954 in China. They found that the highest frequency of such events occurred in the 1950s, but was lowest in the 1990s. Similarly Qian et al. (2002) found high levels of dust activity in the 1950s and a steady decline at Beijing and Baotou thereafter. They suggested that dust storms were twice as prevalent in the 1950s–1970s as they were from the mid-1980s. They attribute this to a reduced meridional temperature gradient, resulting in reduced cyclone frequency in Northern China. Zhao et al. (2003) also attributed the decline in inner Mongolian dust storms from 1961 to 2000 to changes in atmospheric circulation. Likewise, Ding et al. (2005) believed that the sharp decrease in dust storm activity after the mid-1980s occurred concurrently with enhanced geopotential height over the Mongolian Plateau and Middle Siberia as well as an anomalous shift in the phase and intensity of the stationary wave over Eurasia. In contrast, Parungo et al. (1994) attributed the negative trend in dust storm frequency to the planting of a vast belt of forests – 'The Great Green Wall' – across the northern arid lands of China. They asserted that when in the 1980s and 1990s dust storms were rare in Beijing there were not statistically significant changes in wind speed or precipitation. An analysis by Wang et al. (2004b; Fig. 7.15b) confirmed that, for China as a whole, the highest frequencies of dust storms occurred in the 1960s and 1970s, though they recognized that in some regions (such as the Chaidm basin) they were increasing; and they attributed this to localized desertification brought about by human pressures on the land. In the early twenty-first century reduced precipitation and a concomitant decrease in vegetation cover caused a resurgence of dust events (Zhou and Zhai 2004). There also appears to have been greater atmospheric instability, leading to stronger winds and thus more dust storms (Gao et al. 2003b). Finally, Fan and Wang (2004) believed that there was a link between dust weather frequency in northern China and the Antarctic Oscillation, while Liu et al. (2004) suggested that some recent dust storm events have coincided with the occurrence of El Niño events.

Data on dust event trends in Korea are provided by Chun et al. (2001) for the period 1915–1999, though data for 1923–1953 are missing. The number of Asian dust events observed in Seoul appears to have increased sharply since the 1970s (Fig. 7.15c). The years 2000, 2001 and 2003 were especially dusty (Youngsin and Lim 2003). A subsequent study by Chun and Cho (2003) indicated that the period from the 1930s to early 1940s was also characterized by a very high number of dust days. By contrast, records of *Kosa* events at two stations run by the Meteorological Agency of Japan in Yamanashi prefecture show a discernable declining trend since the late 1970s (Fig. 7.15d), despite high year-on-year variability.

7.8   Australia

The Australian continent is marked by considerable variation in dust storm activity from year to year; and agricultural degradation of land surfaces in areas like the Mallee may have had an impact on dust storm frequencies.

The variability of storm frequency across the whole country over the last four decades of the twentieth century is high (Fig. 7.16); and McTainsh et al. (2005) point to the close relationship between years of high dust storm frequency and the occurrence of drought across the continent. Ekström et al. (2004) explore the relationship between dust storm activity and pressure conditions and highlight the importance of variations in the location of the Indian Ocean sub-tropical high, with a more westward displacement of this pressure centre in the Great Australian Bight, allowing cold air to enter the continent, thereby increasing the potential for dust storm activity. Likewise, Leslie and Speer (2005) suggest that the decline in dust storm activity over central eastern Australia, which commenced in the mid-1970s, was due to a decrease in post-frontal south- to south-east winds and that such circulation changes are themselves related to changes in the Pacific decadal Oscillation (PDO).

However, a range of anthropogenic reasons has been forwarded to explain the low frequencies of the 1970s and early 1980s, including a reduction in rabbit numbers, the adoption of minimum tillage techniques and an increase in land cover as a result of the invasion of woody weeds (State of the Environment Advisory Council 1996). However, the occurrence of drought seems likely to be a stronger determinant of dustiness (McTainsh et al. 2005): the 1970s and 1980s being decades with relatively few drought periods. Figure 7.16 indicates that the droughts of 1994–1995 and 2002 were clearly reflected

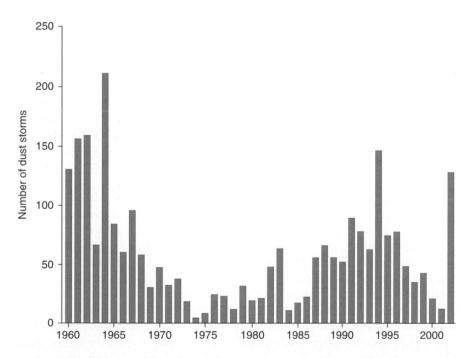

Fig. 7.16. The frequency of dust storms Australia-wide, 1960–2002. Modified after McTainsh et al. (2005, Fig. 5)

in increased dust storm activity, although that of 1982–1983 had a more modest effect in terms of dust storm frequency.

7.9 The Aeolian Environment in a Warmer World

Given the impact that climatic variability in the recent past has had on soil erosion by wind, it is likely that future global warming will have a major impact as well. Changes in precipitation and evapotranspiration rates will probably have a marked impact on the aeolian environment. Rates of deflation, sand and dust entrainment are closely related to soil moisture conditions and the extent of vegetation cover. Areas that are marginal in terms of their stability with respect to aeolian processes will be particularly susceptible; and this has been made evident, for example, through recent studies of the semi-arid portions of the United States (e.g. the High Plains). Repeatedly through the Holocene, they have flipped from states of vegetated stability to states of drought-induced surface instability (Forman et al. 2001). It is likely that many drylands will become drier under global warming, either because of an increasing loss of soil moisture related to higher temperatures, or because of reductions in precipitation inputs. It is also possible that wind velocities may increase.

Indeed, using the output from General Circulation Models, it is apparent that with future global climate change there are likely to be substantial changes in aeolian activity (Muhs and Maat 1993; Stetler and Gaylord 1996), with future dust storm incidence in the High Plains and the Canadian Prairies (Wheaton 1990) being comparable to that of the devastating Dust Bowl years of the 1930s. Modelling studies of southern Africa by Thomas et al. (2005) have suggested that, during the present century, most of the currently largely stable dune surfaces of the Mega-Kalahari will become reactivated and mobile. If this proves to be the case, there will be substantial winnowing of fines from the weathered dune surfaces (see Section 2.3) and an increase in dust storm activity in southern Africa.

At a more localised scale, aeolian processes have already become more active in the past 100 years or so in Iceland as the retreat of glaciers due to climate warming has altered hydrological conditions at glacial margins and in larger sandy areas. The enhanced wind action has buried previously vegetated areas and it is likely that continued glacier retreat will lead to further land degradation (Gisladottir et al. 2005).

However, climate change is not the only factor that will affect future dust storm activity. In addition, not all regions will react in the same way to climate change – some, for example, may become wetter and less dusty, while others may become drier and more dusty. It is also necessary to consider other future environmental changes caused by land use and land cover modifications brought about by human activities (Mahowald and Luo 2003).

7.10   Conclusions

The study of dust storm frequencies over recent decades indicates that different areas show different tendencies. Some regions, such as northern Africa, show an increasing trend of dust emissions, which results from increasing drought, perhaps combined with changing wind velocities and various anthropogenic pressures. The examples of Owens Lake in the United States and the Aral Sea in Central Asia illustrate how man-made desiccation of lake basins can cause dust activity to increase. In contrast, other areas, including parts of the plain lands of North America and parts of China and Australia, show decreasing trends, some of which can be explained by improvements in land management. However, in the first years of the twenty-first century, the downward trend recognised in both Australia and China appears to have been reversed as a result of the return of drought conditions. It is, however, extremely difficult to identify the causes of changes in frequencies with any degree of confidence, because of the complexity of potential factors involved.

8 Dust Storm Control

8.1 Introduction

Various attempts have been made to control the occurrence of dust storms; and these include the array of techniques that have been used for wind erosion control, most of them developed to protect cultivated fields from soil loss (Bennett 1938a, b; Middleton 1990; Riksen et al. 2003a; Sterk 2003; Nordstrom and Hotta 2004). In any particular location, a range of measures is typically employed, as Table 8.1 shows for northern Europe. These techniques are frequently classified into three categories: (a) crop management practices, (b) mechanical tillage operations and (c) vegetative barriers. All of these methods aim to decrease wind speed at the soil surface by increasing surface roughness and/or increasing the threshold velocity that is required to initiate particle movement by wind. Numerous crop management practices, also commonly referred to as agronomic measures, can influence both the detachment and the transport phases of soil particle movement, particularly when combined with good soil management. Mechanical methods, by contrast, effectively do little to prevent soil detachment, but tend to be more effective in preventing soil transport (Morgan 1995).

8.2 Agronomic Measures

Agronomic measures for controlling soil erosion use living vegetation or the residues from harvested crops to protect the soil by acting as non-erodible elements, thereby absorbing the wind's shear stress. When a vegetative cover is sufficiently high and dense to prevent the wind stress on adjacent exposed land exceeding the threshold for particle movement, then the soil will not erode. Roots also help to prevent erosion through their contribution to the mechanical strength of the soil. Maintaining a sufficient vegetative cover is the 'cardinal rule' for controlling wind erosion (Skidmore 1986).

Establishing and maintaining a good ground cover is dependent on the type of crops being grown and the length of time taken to attain a canopy cover of 40–50%. Crops grown in rows, tall tree crops and low-growing crops

Table 8.1. Measures commonly employed to minimise wind erosion risk in northern Europe. After Riksen et al. (2003b)

Aim and timeframe	Technique	Remarks
Techniques that minimise actual risk (short-term)	Autumn-sown varieties	Need to sow before end of October to develop sufficient cover
	Mixed cropping	Second crop remains on the field after main crop is harvested
	Nursing or cover crop	More herbicides needed
	Straw planting	Unsuitable on light sandy soils
	Organic protection layer (e.g. liquid manure, sewage sludge)	Use depends on availability and relevant regulations
	Time of cultivation	Dependent on labour and equipment availability
	Cultivation practices (e.g. minimum tillage, plough and press)	Not suitable for all crop types
Techniques that minimise potential risk (long-term)	Smaller fields	Increase in operational time and costs as well as loss of overall cultivated area
	Change of arable land to alternative use (e.g. permanent pasture, woodland)	Loss of cultivated area, production and farm income
	Marling (increase clay content to 8–10%)	Need suitable material close by
	Wind breaks	High investment costs as well as loss of overall cultivated area

with large leaves offer least protection to the soil surface (Morgan 1995). The simplest way to combine different crops is by rotation, for example by planting a non-commercial crop that will reduce erosion after a cash crop has been harvested. The practice of farming land in narrow strips, on which crop alternates with fallow usually of a legume or grass, is another option. The most effective strips are perpendicular to the prevailing erosive wind direction, but they do provide some protection from winds not perpendicular to the field strip (Skidmore 1986). The strips diminish the wind velocity across the fallow strip, reduce the distance the wind travels over exposed soil and they localise any soil drifting. Strip-cropping demands small fields, however, and thus is not compatible with highly mechanised agriculture; but it provides a useful technique for the smallholder. The maintenance of a crop residue or mulch as a stubble on cropland is recognised as an efficient method for reducing wind erosion losses. The effectiveness of a residue against erosion after harvesting depends on its amount, height, orientation, diameter and density of stalks, as well as its survivability. 'Stubble mulching' is a crop residue management technique that aims to maintain some degree of crop residue on the field surface at all times. The soil is usually tilled, but not to the extent that the field is left 'clean'. The tillage system usually utilises blades or V-shaped sweeps and does not invert the soil (McCalla and Army 1961).

Stubble mulching is a primary erosion control technique used in one form of 'conservation tillage' that does not invert the topsoil and leaves enough crop residue on the field surface after harvest to protect the soil throughout the non-growing season. The farmer plants new seeds among the stalks and debris left from the previous harvest. The method reduces erosion and also reduces farmers' costs since fewer trips with tractor and ploughing equipment are needed through the fields.

The wise management of crop residues is widely used in dryland agriculture in many parts of the world. It is especially valuable in poor countries; e.g. in the Sahel of West Africa, millet mulches of around 2 t ha^{-1} have proved to be highly effective (Bielders et al. 2001). In an experiment to determine the loss of topsoil prevented by millet mulch in Niger, Michels et al. (1995) found a relative difference in surface elevation of 33 mm after just one year between bare millet plots and those spread with 2000 kg ha^{-1} of mulch, as a result of wind erosion and sediment deposition.

The protective properties of crop residues are greatest when the material is left standing rather than being flattened, but standing residue is not always agronomically acceptable. Millet stalks in the Sahel may harbour crop-damaging stem borers (*Acigona ignefusalis* Hampson) when left standing, but the larvae of this pest usually perish – by heat at the soil surface – if the stalks are cut down after harvest, so reducing the risk of infestation during the next cropping season (Ndoye and Gahukar 1987).

In other parts of the world, however, fallen crop residues can still provide a good habitat for insects and weeds. In countries where pesticides are afford-able, this problem can be overcome with chemical applications, but with the concomitant hazards of off-field pollution, killing of non-target species and development of resistance to the chemical used. Where pesticides are not used, the insects and weeds can combine to reduce yields by eating crops and competing for soil nutrients.

Some of these difficulties can be avoided by using other forms of stabiliser. Dung, which is widely used in subsistence agriculture because of its fertilis-ing properties, also provides effective protection to the soil against particle creep and saltation – initiators of suspension – even at a very low level of cover (de Rouw and Rajot 2004). Rock fragments are another widely accepted stabiliser. Pebble and gravel mulches have been used by farmers in north-west China for more than 300 years, to dampen down soil erosion and to trap dust carried by the wind (Li et al. 2001). The accumulation of dust may sup-ply valuable additional nutrients to gravel-mulched fields (Li and Liu 2003). In some countries, sandy soils can be stabilised by the addition of clay to the soil. This process is often called marling; and it reduces erosion risk by increasing aggregate stability.

A range of synthetic materials have also been evaluated for their applicability to wind erosion control (e.g. Armbrust and Dickerson 1971). Among substances used are polyvinylacetate (PVA) emulsions and polyacrylamides (PAM) sprayed onto the soil surface. These can provide temporary protection for high-value crops but are too expensive for low-value crops. Polymers are applicable to the control of saline dust blowing from tailing ponds (Fuller and Marsden 2004). Some stabilisers have been found to meet the essential criteria for soil surface stabilisers (Armbrust and Lyles 1975):

(a) One hundred per cent of the soil must be covered.
(b) Stabilisers must not adversely affect plant growth or emergence.
(c) Erosion must be prevented initially and reduced throughout the period of severe erosion hazard.
(d) The stabiliser must be easily applied and without special equipment.
(e) The cost must be low enough for profitable use.

8.3   Soil Management

Soil management techniques focus on ways of preparing the soil to promote good vegetative growth and to improve soil structure in order to increase resistance to erosion. Applying organic matter is a form of soil management that can decrease soil erodibility as well as enhance its fertility, but most soil management methods pertaining to erosion control are concerned with different forms of tillage.

Tillage is an essential part of farming, providing a suitable seed bed for plant growth and helping to control weeds, but the dangers of inappropriate tillage have been illustrated in the Maghreb of North Africa, the United States Great Plains and the Virgin Lands scheme of the former Soviet Union (see Chapter 7). Excessive tillage, particularly of light-textured soils, breaks soil clods, reduces surface roughness and exposes soil to wind action, particularly if soil-overturning binds stubble into the soil, thus reducing mulch coverage. To overcome this destruction of structure in non-cohesive soils, tillage operations must be restricted. This may be by reducing the number of passes over a field by combining as many operations into one pass as possible, such as in mulch tillage or minimum tillage, or by strip-zone tillage where operations are concentrated only as rows where the plants grow, leaving the inter-row areas untilled (Schwab et al. 1966; see Table 8.2).

The effects of various forms of conservation tillage on erosion rates, soil conditions and crop yields has been the subject of many studies in recent years (see, for example, Merrill et al. 1999) and the results show the success of the system to be highly soil-specific and also to depend on how well weeds, pests and diseases are controlled (Morgan 1995). To give one example, significant differences in dust production from field experiments in semi-arid

Table 8.2. Tillage practices used for soil conservation. After Schwab et al. (1966)

Practice	Description
Conventional	Standard practice of ploughing with disc or mouldboard plough, one or more disc harrowings, spike-tooth harrowing and surface planting
Strip or zone tillage	Preparation of seed bed by conditioning the soil along narrow strips in and adjacent to the seed rows, leaving the intervening soil areas untilled
Mulch tillage	Practice that leaves a large percentage of residual material (leaves, stalks, crowns, roots) on or near the surface as a protective mulch
Minimum tillage	Preparation of seed bed with minimal disturbance; use of chemicals to kill existing vegetation, followed by tillage to open only a narrow seed band to receive the seed; weed control by herbicides

north-east Spain were detected by López et al. (1998) when conventional tillage operations (mouldboard ploughing) were compared to reduced tillage (chisel ploughing). Reduced tillage produced a smaller wind-erodible fraction at the soil surface and a greater percentage of soil cover with crop residues and clods, resulting in lower values of vertical dust flux.

The practice of no-tillage agriculture, in which drilling is carried out directly into the stubble of the previous crop, has been found to show great promise (e.g. Phillips et al. 1980). It reduces labour costs and soil and moisture losses, enhances soil organic matter content, and maintains good structure. Schmidt and Triplett (1967, quoted in Phillips et al. 1980) showed soil erosion loss from a no-tillage field of corn in Ohio to be 4.5 t ha^{-1} during a severe windstorm as compared to a conventionally planted cornfield that lost 291 t ha^{-1}. In Nebraska, the use of no-tillage and herbicides to control wind erosion resulted in less weed growth, higher soil moisture storage and higher grain yields than conventional tillage over a 6-year period (Wicks and Smith 1973). No-tillage has been embraced by agrochemical companies because it requires heavier doses of pesticides, but this in itself is not necessarily desirable due to possible increases in off-field pollution (Risser 1985). Plant residues on no-tillage fields may lower soil temperatures by as much as 6 °C at 25 mm depth in spring, which can delay spring plantings in central and northern North America where soil temperatures are below those needed for optimal growth, but in the tropics this effect may be useful where soil temperatures are frequently above the optimum for maximum plant growth (Phillips et al. 1980). Nevertheless, experience in north-western India suggests that, because of the low organic matter content of sandy soils in arid areas, they become compacted with no-tillage systems, which seriously reduces the growth and yield of crops (Gupta et al. 1983).

A technique developed by the United States Department of Agriculture in Arizona, specifically for grassland re-vegetation, involves the 'firming' and 'shaping' of the land surface. 'Land imprinting' refines the function in nature in which hoof prints from grazing ungulates perform the role of seed bed preparation by holding rainwater for soil infiltration and thereby allowing 'nature-irrigated' germination if a seed is present (Anderson 1987). The imprinting machine consists of a single rolling cylinder, the only moving part, attached to a pulling frame. The imprints on the soil are made by angle irons welded to the cylinder; their configuration can be adapted to specific site conditions. The design is so simple that the machine can be made in any sophisticated welding workshop anywhere in the world.

8.4   Mechanical Methods

Mechanical approaches to wind erosion control manipulate the surface topography in order to control the flow of the wind. Such techniques include the creation of barriers to wind flow such as fences and windbreaks (known as shelterbelts when composed of living plants) and altering surface topography, such as by ploughing furrows.

Barriers to wind flow aid erosion control by decreasing surface shear stress in their lee and by acting as a trap to moving particles, although barriers also create turbulence in their lee which can reduce their effective protection. Their efficiency in terms of reduction of wind velocity and turbulence intensity is determined by a range of factors, including barrier porosity (dependent on plant spacing, stalk and leaf width), porosity distribution, shape, height, orientation, width and spacing. The most efficient barrier is semi-permeable because, although its velocity reduction is less than for an impermeable fence, the amount of eddies and turbulence in its lee are reduced (Cooke et al. 1982). In the same way, windbreaks and shelterbelts should be designed to optimise the interaction between height, density, porosity, shape and width of the plant barrier (Cornelis and Gabriels 2005). A barrier oriented perpendicular to winds predominantly from a single direction will decrease wind erosion forces by more than 50% from the barrier leeward to 20 times its height, the decrease being greater at shorter distances from the barrier (Skidmore 1986). In situations where erosive winds come from several directions, grid or herringbone layouts provide better all-round protection.

Numerous other benefits to crops can also often be associated with the establishment of windbreaks. These include increased soil and air temperatures, improved plant water relations and irrigation efficiency, reduced pest and disease problems and an extended growing season in sheltered areas, resulting in increased crop development, earlier crop maturity and market advantage (Hodges and Brandle 1996). Many studies have also documented yield increases in crops grown behind windbreaks, although effects on yield can vary greatly between crops, situations and seasons (Baldwin 1988).

Shelterbelts are composed of a range of shrubs, tall-growing crops and grasses, besides the more conventional tree windbreak. They may be planted, left as remnants of formerly forested areas, or allowed to grow naturally in fencerows after tractors have cleared fields. However, most barrier systems occupy space that would otherwise be used for crops. Perennial barriers grow slowly, can be difficult to establish and compete with crops for water and plant nutrients (Dickerson et al. 1976; Lyles et al. 1983). Thus the net effects of living barrier systems must be weighed against possible adverse effects on yields (e.g. Frank et al. 1977). Some of these difficulties can be avoided by using artificial barriers, such as stone walls, wood or fabric fences, but the economic costs of materials and labour to construct them often restrict their use to high-value crops (Tibke 1988).

The ploughing of ridges is a common anti-erosion measure that acts to roughen the soil surface and thus reduce the average wind velocity for some distance above the ground. Ridges also trap entrained particles on their leeward sides (Chepil and Milne 1941). Tillage to produce ridges across the path of the erosive wind is usually carried out by chisel and is successfully used temporarily to control wind erosion in an emergency (Woodruff et al. 1957). Farmers of sandy soils in the Midland counties of England employ a version of ridge and furrow tillage to control wind erosion on land devoted to sugar beet (Morgan 1995). The Glassford system ploughs soil that is moist but not wet to produce ridges and furrows and immediately the furrows are rolled. The operation is carried out in January and the resulting furrowed and ridged surface remains stable throughout the spring blowing period. However, in poorer farming regions such as the Sahel, where mechanical measures depend on animal traction, the technique is not so widely used. Also, because of the sandy soils, ridges and furrows are short-lived, being broken down during rain storms (Bielders et al. 2000).

8.5   Miscellaneous Methods to Reduce Dust Emissions

Fugitive dust emissions also warrant the use of various other suppression techniques, including the application of water by means of trucks, hoses and/or sprinklers prior to conducting any activities that might disturb the surface. Such short-term control techniques may be complemented by the cessation of activities at times of high wind velocity. Surfaces can be stabilised for longer periods by paving dirt tracks or applying dust suppressant chemicals.

The stabilisation of desiccated lake beds is a particularly important issue with respect to locations like Owen's Lake in California. In an ideal world, stream flows that are currently being diverted would be returned to the basin. Given that this is unlikely to be possible, other techniques have been experimentally trialled in the Owen's Valley (Gill and Cahill 1992), including sand fences to catch coarse particles, chemical surfactants, the spreading of gravel, mechanical compaction, sprinkler irrigation and re-vegetation.

9 Quaternary Dust Loadings

9.1 Introduction

At certain times during the Quaternary, such as the Last Glacial Maximum (LGM) at around $18-20\times10^3$ years ago (Mahowald et al. 1999), the world was a very dusty place. This is indicated by its extensive deposits of loess, the presence of large amounts of aeolian dust in ocean, lake and peat bog core sediments, the existence of quantities of dust found in ice cores drilled from the polar regions and elsewhere and even the accumulations of desert dust in speleothems. These natural archives have been intensively studied for their palaeoenvironmental significance (e.g. Muhs and Bettis 2000; Shichang et al. 2001; Pichevin et al. 2005). The enhanced dustiness they have accumulated, especially during cold glacial periods, may relate to a larger sediment source (e.g. areas of glacial outwash), changes in wind characteristics both in prox-imity to ice caps and in the trade-wind zone (Ruddiman 1997) and the expan-sion of low-latitude deserts. It would be simplistic to attribute all cases of higher dust activity to greater aridity in source regions for, as Nilson and Lehmkuhl (2001) point out, this is but one factor, albeit important. Also important are changes in the trajectories of the major dust-transporting wind systems, changes in the strength of winds in source regions, the balance between wet and dry deposition (which may determine the distance of dust transport), the degree of exposure of continental shelves in response to sea-level changes and the presence of suitable vegetation to trap dust on land. It is possible that increased dust loadings during the LGM were not only a prod-uct of climatic change but also a contributory factor to that change; and this is something that is now being built into climatic models (e.g. Overpeck et al. 1996; Mahowald et al. 1999).

9.2 Ocean Cores

It is possible to obtain a long-term measure of dust additions to the oceans by undertaking studies of the sedimentology of deep-sea cores (Rea 1994). Working in the Arabian Sea, Clemens and Prell (1990) found a positive correlation between global ice volume (as indicated by the marine O^{18}/O^{16}

record) and the accumulation rate and sediment size of dust material. Kolla and Biscaye (1977) confirmed this picture for a larger area of the Indian Ocean and indicated that large dust inputs came off Arabia and Australia during the last glacial. On the basis of cores from the Arabian Sea, Sirocko et al. (1991) suggested that dust additions were around 60% higher during glacials than in post-glacial times, though there was a clear 'spike' of enhanced dust activity at around 4000 years BP associated with a severe arid phase that has been implicated in the decline of the Akkadian empire (Cullen et al. 2000). Jung et al. (2004) also report on Holocene dust trends in the Arabian Sea and suggest that dry, dusty conditions were established by 3800 years BP.

Pourmand et al. (2004) refined this further and showed that high dust fluxes in the Middle East occurred during cold phases such as the Younger Dryas, Heinrich events 1–7 and cold Dansgaard–Oeschger stadials. They attributed this to a weakened south-west monsoon and strengthened north-westerlies from the Arabian Peninsula and Mesopotamia. Similarly, a core from the Alboran Sea in the western Mediterranean indicated an increase in dust activity during Dansgaard–Oschger stadials and Heinrich events (Moreno et al. 2002).

There is particularly clear evidence for increased dust inputs at the time of the LGM, at around 18×10^3 years ago (Fig. 9.1). In the Atlantic offshore from the Sahara, the amount of dust transported into the Ocean was augmented by a factor of 2.5 (Tetzlaff et al. 1989, p. 198). Australia contributed three times more dust to the south-west Pacific Ocean at that time (Hess and McTainsh 1999) and increased dust loadings to the ocean may have stimulated increases in planktonic productivity on the South Australian continental margin (Gingele and De Deckker 2005). Dust fluxes appear generally to have been two to four times higher than at present (Kolla et al. 1979; Sarnthein and Koopman 1980; Tetzlaff and Peters 1986; Chamley 1988; Grousset et al. 1998). By contrast, they appear to have been very low during the 'African Humid Period' (AHP). From 14.8×10^3 years ago to 5.5×10^3 years ago, the mass flux off Cape Blanc was reduced by 47% (DeMenocal et al. 2000). This is confirmed by analyses of the mineral magnetics record from Lake Bosumtwi in Ghana, which suggest a high dust flux during the last glacial period and a great reduction during the AHP (Peck et al. 2004).

The causes of high dust fluxes during glacial phases include reductions in precipitation. However, changes in the strength of the north-easterly trades may also have been a major contributory factor in some areas in the northern hemisphere (Ruddiman 1997; Grousset et al. 1998; Moreno et al. 2001; Abouchami and Zabel 2003) and various studies have been made of wind-transported materials (including phytoliths, diatoms deflated from desiccated lakes and also grain sizes) to plot wind strength changes over extended periods (e.g. Hooghiemstra 1989; Stabell 1989; Abrantes 2003; Pichevin et al. 2005). However, evidence for stronger winds during the LGM is not universal, with Hesse and McTainsh (1999) arguing that this was not a factor in the higher dust loadings in the Tasman Sea at that time.

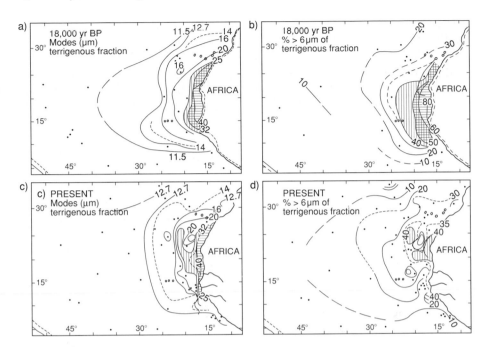

Fig. 9.1. Dust in the Atlantic off the Sahara at 18 000 years BP (the glacial maximum) and 6000 years BP (mid-Holocene), as revealed by ocean core sediments. a) Distribution of modal grain sizes of terrigenous silt (>6 μm, carbonate, opal-free) at 18 000 BP. b) Distribution of percentage silt (>6 μm, carbonate, opal-free) at 18 000 BP. c) Distribution of modal grain sizes of terrigenous silt (>6 μm, carbonate, opal-free) in surface sediments. d) Distribution of percentage terrigenous silt (>6 μm). Modified after Sarnthein and Koopmann (1980, Figs. 2, 3, 5, 6)

Bozzano et al. (2002), on the basis of their analysis of an ocean core off Morocco, found a correlation between dust supply and precessional minima in the earth's orbit. They argued that enhanced precession-driven solar radiation in the boreal summer would have increased seasonal temperature contrasts, which in turn amplified atmospheric turbulence and stimulated storminess. In other words, they believe that a crucial control of dust storm activity is not simply aridity, but the occurrence of meteorological events that can raise dust from desert surfaces.

Cores from the Japan Sea (Irino et al. 2003) show the importance of dust deposition at the maximum of the LGM (Fig. 9.2). Both the amount of silt being deposited and its modal size indicate an intensification of dust supply at that time. In the mid-latitude North Pacific, which is also supplied with dusts from Central Asia, dust deposition maxima during the last 200×10³ years occurred in OIS 4 to latest OIS 5 and in the middle of OIS 6 (Kawahata et al. 2000). These were seen as times of reduced precipitation during the summer monsoon and strengthened wind speeds during the winter monsoon.

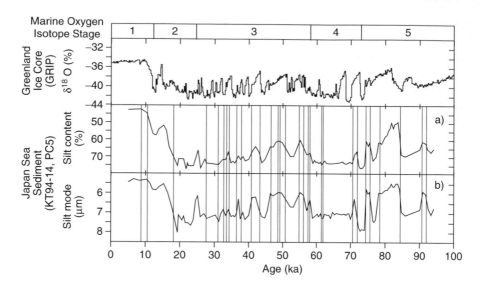

Fig. 9.2. Temporal variations of aeolian dust (silt) content a) and grain size b) in core KT94-15-PC5 recovered from the Japan Sea. Oxygen isotope variations from GRIP ice core are also shown above for comparison. Modified after Irino et al. (2003, Fig. 2)

At a longer time-scale, there is some evidence the dust activity increased as climate deteriorated during the late Tertiary. In the Atlantic off West Africa, Pokras (1989) found clear evidence for increased terrigenous lithogenic input at $2.3-2.5\times10^6$ years ago, while Schramm (1989) found that the largest increases in mass accumulation rates in the North Pacific occurred between 2×10^6 and 3×10^6 years ago. This coincides broadly with the initiation of northern hemisphere glaciation. However, no such link has been identified in the southern Pacific Ocean (Rea 1989).The lengthiest analysis of dust deposition in the oceans was undertaken by Leinen and Heath (1981) on sediments of the central part of the North Pacific. They demonstrated that there were low rates of dust deposition $50-25\times10^6$ years ago. This they believe reflects the temperate, humid environment that was seemingly characteristic of the early Tertiary and the lack of vigorous atmospheric circulation at that time. From 25×10^6 to 7×10^6 years ago, the rate of aeolian accumulation on the ocean floor increased, but it became greatly accelerated from 7×10^6 to 3×10^6 years ago. However, although there is thus an indication that aeolian processes were becoming increasingly important as the Tertiary progressed, it was around 2.5×10^6 years ago that there occurred the most dramatic increase in aeolian sedimentation. This accompanied the onset of northern hemisphere glaciation.

Deposition of dust in the North Pacific occurred before the oldest preserved Asian loess formed, but isotopic studies indicate it came from the basins of Central Asia. Over the past 12×10^6 years, however, the dust flux to the North Pacific has increased by more than an order of magnitude, documenting a substantial drying of Central Asia (Pettke et al. 2000).

The analysis of deep-sea cores in the North Atlantic provides a picture of long-term changes in dust supply and aeolian activity in the Sahara. Some dust dates back to the early Cretaceous (Lever and McCave 1983) and aeolian dust is present in Neogene sediments (Sarnthein et al. 1982). However, aeolian activity appears to become more pronounced in the late Tertiary. As Stein (1985, pp. 312–313) reported: "Distinct maxima of aeolian mass accumulation rates and a coarsening of grain size are observed in the latest Miocene, between 6 and 5 Ma and in the Late Pliocene and Quaternary, in the last 2.5 million years". They attribute this to both a decrease in precipitation in the Sahara and to an intensified atmospheric circulation. The latter was probably caused by an increased temperature gradient between the North Pole and the Equator due to an expansion in the area of northern hemisphere glaciation. From about 2.5×10^6 to 2.8×10^6 years ago, the great tropical inland lakes of the Sahara began to dry out; and this is more or less contemporaneous with the time of onset of mid-latitude glaciation. High dust loadings were a feature of the Pleistocene (Pokras 1989). Mean late-Pleistocene dust inputs were two to five times higher than the pre-2.8×10^6 year values (DeMenocal 1995).

In the Mediterranean basin, which derives much of its dust load from the Sahara, Larrasoaña et al. (2003) analysed a core from the seabed south of Cyprus, using its haematite content as a proxy for dust. It covered a period of three million years. They found that, throughout that time, dust flux minima occurred when the African summer monsoon attained a northerly position during times of insolation minima. This, they argued, increased the vegetation cover and soil moisture levels, thereby dampening down dust activity in the Saharan source regions.

9.3 Dust Deposition as Recorded in Ice Cores

Another major source of long-term information on rates of dust accretion is the record preserved in long ice cores retrieved either from the polar ice caps or from high-altitude ice domes at lower altitudes. Indeed, observations of dust in polar ice cores has done much to establish the reality of abrupt climate changes in the Quaternary and dust has been described as climate's 'Rosetta Stone' (Broecker 2002).

Because they are generally far removed from source areas, the actual rates of accumulation of dust in ice cores are generally low, but studies of variations in micro-particle concentrations with depth do provide insights into the relative dust loadings of the atmosphere in the last glacial and during the course of the Holocene. Thompson and Mosley-Thompson (1981) drew together a lot of the material that was published at the time they wrote and pointed to the great differences in micro-particle concentrations between the Late Glacial and the Post-Glacial. The ratio for the Dome C ice core (E. Antarctica) was 6:1, for the Byrd Station (W. Antarctica) 3:1, and for Camp Century (Greenland) 12:1.

Briat et al. (1982) maintained that, at Dome C, there was an increase in micro-particle concentrations by a factor of 10–20 during the last glacial stage; and they explain this by a large input of continental dust. The Dunde ice core from High Asia (Thompson et al. 1990) also shows very high dust loadings in the Late Glacial and a very sudden fall off at the transition to the Holocene. Within the last glaciation, dust activity both in Europe and in Greenland appears to have varied in response to millennial-scale climatic events (Dansgaard–Oeschger Events and Bond Cycles; Rousseau et al. 2002).

These early results are confirmed by the more recent study of the Epica and Vostok cores from Antarctica (Delmonte et al. 2004a; Fig. 9.3). In the Epica core (Fig. 9.4), the dust flux rose by a factor of ca. 25, ca. 20 and ca. 12 in Glacial Stages 2, 4 and 6 compared to interglacial periods (the Holocene and OIS Stage 5.5). Delmonte et al. (2004b) found in the Dome B, Vostok and Komsomolskaia cores that, during the LGM, dust concentrations were

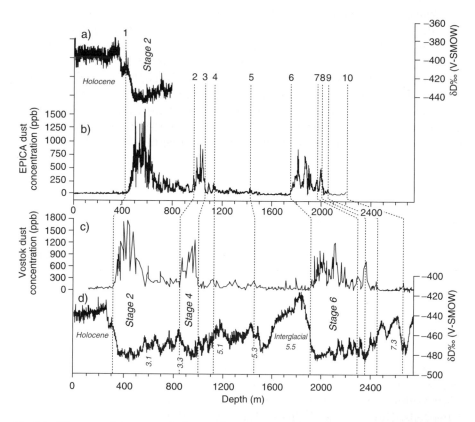

Fig. 9.3. Climate and dust records from EPICA Dome C and Vostok ice cores. a) EPICA deu-terium record. b) EPICA dust concentration record (ppb) to 2201 m depth. c) Vostok dust con-centration record (ppb) to 2670 m depth. d) Vostok deuterium record for the past ca. 220 000 years, with the major climatic stages indicated. The *dashed lines* linking EPICA and Vostok ice cores identify ten common dust events (*1–10*). Modified after Delmonte et al. (2004a, Fig. 2)

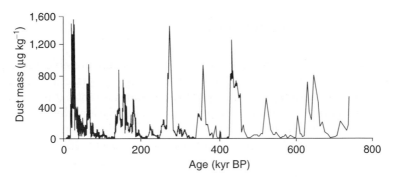

Fig. 9.4. Dust mass from EPICA Dome C core, Antarctica over more than 700 000 years. Modified after EPICA community members (2004, Fig. 2D)

between 730 ppb and 854 ppb, whereas during the Antarctic Cold Reversal (14.5–12.2×10³ years BP) they had fallen to 25–46 ppb and, from 12.1×10³ to 10×10³ years BP, they were between 7 ppb and 18 ppb. Isotopic studies suggest that the bulk of the dust was derived from Patagonia and the Pampas of Argentina (see also Iriondo 2000). In the case of Greenland, a prime source of dust in cold phases was East Asia (Svensson et al. 2000). Broecker (2002) suggests that the increase in dust production and deposition in glacial times can be attributed to the steepened temperature gradients and associated aeolian activity related to the equatorward extension of continental glaciers and sea ice. However, changes in the hydrological and vegetative state of source regions will also have been very important (Werner et al. 2002).

Studies of dust in ice cores can also be applied to recent decades. The North GRIP core in Greenland indicates that, in the late 1990s, east Asia was a major source and the provenance in spring/summer was the Taklamakan Desert (Bory et al. 2002). In contrast, the GISP2 core from Greenland shows dust that originated in the United States during the 1930s Dust Bowl (Donarummo et al. 2003). An ice core from near Mount Everest shows a series of intense dust periods during the past 200 years (Kang et al. 2001), particularly in the 1830s to 1840s and in the 1890s to 1920s. A core from Dasuopu, Tibet, shows intense dust accumulation from 1790 AD to 1796 AD, a time of severe drought in India.

Although studies of cores from the Atlantic, Indian and Pacific Oceans and from polar ice tend to show the importance of dust accumulation during cold phases, this is not a universal picture. Thus, areas that were covered in snow and had extensive freshwater lakes in glacial phases might have generated limited amounts of dust; and this is the explanation provided by Thompson et al. (1998), who found that LGM ice from the Sajama ice cap in the high mountains of Bolivia contains eight times less dust than the Holocene ice. In contrast, ice from the Late Glacial at Huascarán, Peru, indicates it was a time of extreme dustiness because of high winds and drier surface conditions (Thompson et al. 1995).

9.4 Loess Accumulation Rates

By measuring and dating loess sections, it has been possible to estimate the rate at which loess accumulated on land during the Quaternary (see Table 9.1). The presented data may somewhat underestimate total dust fluxes into an area because, even at times of rapid loess accumulation, there would have been concurrent losses of material as a result of fluvial and mass-movement processes. Solution and compaction may also have occurred.

The data in Table 9.1 show a range of values between 22 mm and 4000 mm per 1000 years. Pye (1987, p. 265) believes that, at the LGM, loess was probably accumulating at a rate of between 500 mm and 3000 mm per 1000 years and suggests that: "Dust-blowing on this scale was possibly unparalleled in previous Earth History". By contrast, he suggests that: "During the Holocene, dust deposition rates in most parts of the world have been too low for significant thicknesses of loess to accumulate, although aeolian additions to soils and ocean sediments have been significant". Pye also hypothesises that rates of loess accumulation showed a tendency to increase during the course of the Quaternary. Average loess accumulation rates in China, Central Asia and Europe were of the order of 20–60 mm per 1000 years during Matuyama time (early Pleistocene) and of the order of 90–260 mm per 1000 years during the Brunhes epoch (post-0.78×10^6 years ago). He also points out that these long-term average rates disguise the fact that rates of loess deposition were one to two orders of magnitude higher during Pleistocene cold phases and were one or two orders of magnitude lower during the warmer interglacial phases when pedogenesis predominated.

A very detailed analysis of loess accumulation rates in China is provided by Kohfeld and Harrison (2003). They indicate that in the glacial phases

Table 9.1. Loess accumulation rates for the late Pleistocene. From various sources in Pye (1987, Table 9.6) and Gerson and Amit (1987)

Location	Accumulation rate (mm per 1000 years)
Negev (Israel)	70–150
Mississippi Valley (USA)	700–4000
Uzbekistan	50–450
Tajikistan	60–290
Lanzhou (China)	250–260
Luochaun (China)	50–70
Czechoslovakia	90
Austria	22
Poland	750
New Zealand	2000

(e.g. OIS 2) aeolian mass accumulation rates were ca. 310 g m^{-2} year^{-1} compared to 65 g m^{-2} year^{-1} for an interglacial stage (e.g. OIS 5) – a 4.8× increase. A comparable exercise was carried out for Europe by Frechen et al. (2003). They found large regional differences in accumulation rates but suggested that, along the Rhine and in eastern Europe, rates were from 800–3200 g m^{-2} year^{-1} in OIS 2. Loess accumulation rates over much of the United States during the LGM were also high, being around 3000 g m^{-2} year^{-1} for mid-continental North America (Bettis et al. 2003). From 18×10^3 years ago to 14×10^3 years ago, rates of accumulation in Nebraska were remarkable, ranging from 11 500 g m^{-2} year^{-1} to 3500 g m^{-2} year^{-1} (Roberts et al. 2003).

Further details from the large number of studies devoted to loess are covered in Chapter 10.

10   Loess

10.1   Introduction

Loess has been the subject of an enormous literature, ever since Charles Lyell (1834) drew attention to the loamy deposits of the Rhine Valley in Germany. Many theories have been advanced to explain loess formation; and Smalley (1975) provides excerpts from the early literature and a commentary to go with them. It was, however, Ferdinand von Richthofen (1882, pp. 297–298) who cogently argued that these intriguing deposits probably had an aeolian origin and that they were produced by dust storms transporting silts from deserts and depositing them on desert margins:

"In regions where the rains are equally distributed through the year, little dust is formed, and the rate of growth of the soil covered with vegetation will be exceedingly small. But where a dry season alternates with a rainy season, the amount of dust which is put in motion and distributed through atmospheric agency can reach enormous proportions, as witnessed by the dust storms which in Central Asia and Northern China eclipse the sun for days in succession. A fine yellow sediment of measurable thickness is deposited after every storm over large extents of country. Where this dust falls on barren ground, it is carried away by the next wind; but where it falls on vegetation, its migration is stopped.

"In rainless deserts the wind will gradually remove every particle of fine-grained matter from the soil, though a new supply of this may constantly be provided by the action of sandblast. The sediments of desiccated lakes, the soil which is laid bare by the retiring of the sea, the materials which are carried down by periodical torrents from glaciated regions to desert depressions, the particles which on every free surface of rock are loosened by constant decay – all these will be turned over and over again by the wind . . ."

While it is true that the silt carried by the wind may result from a wide range of processes, including glacial grinding (see Section 2.2), and that silts may be re-worked and modified by pedological processes, mass movements and fluvial activity, the case for an aeolian role in loess formation is overwhelming.

Loess is largely non-stratified and non-consolidated silt, containing some clay, sand and carbonate (Smalley and Vita-Finzi 1968). It is markedly finer than aeolian sand. Many parts of the world possess long sequences of loess and palaeosols (Rutter et al. 2003) and these provide a major source of palaeoen-vironmental information that can be correlated with that obtained from ocean cores. It consists chiefly of quartz, feldspar, mica, clay minerals and carbonate grains in varying proportions; and Table 10.1 gives some details of major ele-ment geochemistry of unweathered loess. The grain size distribution of typi-cal loess shows a pronounced mode in the range 20–40 µm and is generally positively skewed towards the finer sizes. It can, however, sometimes have a sand content of over 20%, in which case it is termed 'sandy loess', or a clay content in excess of 20%, in which case it is termed 'clayey loess' (Pye 1987, p. 199). Grain size depends on distance from source, formative wind velocities and the granulometry of the materials from which it is derived.

Loess is present in the ancient stratigraphic record, as for example in the Palaeozoic beds of Utah (Soreghan et al. 2002), but in this section we concen-trate primarily on the great Quaternary loess accumulations, which cover as much as 10% of the Earth's land surface (Muhs et al. 2004). Over vast areas (at least 1.6×10^6 km^2 in North America and 1.8×10^6 km^2 in Europe), these blanket the pre-existing relief and, in Tajikistan, these accumulations have been recorded as reaching a thickness of up to 200 m (Frechen and Dodonov 1998). In the Missouri Valley of Kansas, the loess may be 30 m thick. European Russia has sustained thicknesses, often 10–30 m and reaching over 100 m in places, while in New Zealand, on the plains of the South Island, thicknesses reach 18 m. Loess profiles thicker than 50 m are known from boreholes in the Pampas of Argentina (Kröhling 2003).

Loess is known from some high-latitude regions, including Greenland, Alaska (Muhs et al. 2004), Spitzbergen, Siberia (Chlachula 2003) and Antarctica (Seppälä 2004). Loess has also been recorded from various desert regions (Table 10.2). In Arabia, Australia and Africa, where glaciation was relatively slight, loess is much less well developed, though an increasing num-ber of deposits in these regions is now becoming evident. Of all the world's loess deposits, those of China are undoubtedly the most impressive for their extent and thickness, which near Lanzhou is 300–500 m.

The distribution of loess in North America is now well known; and the main areas in the United States include southern Idaho, eastern Washington,

Table 10.1. Major element geochemistry of unweathered loess in comparison to dust (%)

Component	Loess[a]	Dust[b]
SiO_2	63.80 (53.1–82.03)	59.9
Al_2O_3	10.41 (7.52–16.13)	14.13
Fe_2O_3	3.75 (2.77–5.10)	6.85
MgO	2.34 (0.65–4.53)	2.60
CaO	6.99 (0.61–13.56)	3.94

[a] Mean and range (in brackets) based on 15 samples in Pye (1987, Table 9.2)
[b] Based on data in Table 6.9

Table 10.2. Examples of peridesert loess

Location	Reference (s)
Matmata, Tunisia	Coudé-Gaussen et al. (1982), Dearing et al. (1996, 2001)
Namib	Blümel (1982)
Northern Nigeria	McTainsh (1987)
Eastern Afghanistan	Pias (1971)
Potwar, Pakistan	Rendell (1984)
Negev	Yaalon and Dan (1974)
Syria	Rösner (1989)
Iran	Lateef (1988), Okhravi and Amini (2001), Kehl et al. (2005)
Bahrain	Doornkamp et al. (1980)
Yemen	Nettleton and Chadwick (1996), Coque-Delhuille and Gentelle (1998)
United Arab Emirates	Goudie et al. (2000)
Saudi Arabia	Al-Harthi and Bankher (1999)
Peru	Eitel et al. (2005)

north-eastern Oregon and, even more important, a great belt from the Rocky Mountains across the Great Plains and the Central Lowland into western Pennsylvania. Loess is less prominent in the eastern United States as relief, climatic conditions for deflation and the nature of outwash materials seem to have been less favourable than in the Missouri–Mississippi region. There are at least four middle-to-late Quaternary loess units in the High Plains, which from oldest to youngest are the Loveland Loess (Illinoian glacial), the Gilman Canyon Formation (mid- to late Wisconsinian), the Peoria Loess (late Wisconsinian) and the Bignell Loess (Holocene; Pye et al. 1995; Muhs et al. 1999). The loess deposits of the United States have recently been reviewed by Bettis et al. (2003; Fig. 10.1), who suggest that the Last Glacial (Peoria) loess is probably the thickest in the world, being more than 48 m thick in parts of Nebraska and 41 m thick in western Iowa. Some of the Peoria loess, including than in Nebraska, may not be glaciogenic, having been transported by westerly to northerly winds from parts of the Great Plains not directly influenced by the Laurentide ice sheet or alpine glaciers (Mason 2001). However, this has been a matter of some controversy, for Winspear and Pye (1995) favoured a more glacial explanation for the Peoria Loess in Nebraska. Some of the loess in the Great Plains (the Bignell Loess) is of Holocene age (Mason and Kuzila 2000; Mason et al. 2003; Jacobs and Mason 2005). Miao et al. (2005) believe that much of the Holocene loess, most of which dates from 9000–10 000 years to 6500 years ago, was produced in dry phases as a result of the winnowing of dune fields.

In South America, where the Pampas of Argentina and Uruguay has thick deposits, a combination of semi-arid and arid conditions in the Andes

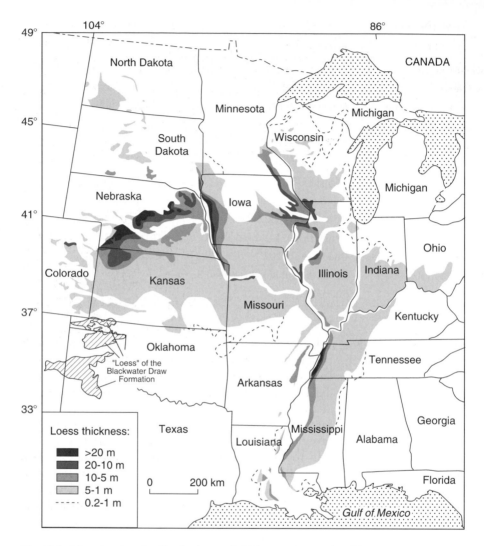

Fig. 10.1. Map showing the distribution and thickness of Last Glacial loess (Peoria Loess) in mid-continental USA (Central Lowland and Great Plains physiographic provinces). Modified after Bettis et al. (2003, Fig. 2)

rain-shadow, combined with glacial outwash from those mountains, created near ideal conditions (Zarate 2003). The Argentinian loess region is the most extensive in the Southern Hemisphere, covering 1.1×10^6 km^2 between 20° S and 40° S. Zinck and Sayago (2001) described a 42-m thick loess – palaeosol sequence of Late Pleistocene age from north-west Argentina, though generally thicknesses are less than this. Much of the loess was laid down in the Late Pleistocene during the Last Glacial Maximum, but some deposition has also occurred in the Holocene. There is isotopic evidence that some of the loess contains a substantial amount of dust derived from volcanic sources (Sayago et al. 2001; Smith et al. 2003), but multiple geomorphological sources have also been proposed, including the Argentinian continental shelf, the Paraná River Basin, the Pampean Hills, the Altiplano-Puna Plateau and glaciofluvial deposits from Mendoza, Neuquen and Rio Negro. Mantles of aeolian silt and loess are known from other parts of South America, including the Orinoco Llanos of Colombia and Venezuela, north-east Brazil, the central valley of Chile and southern Peru (Iriondo 1997; Iriondo and Kröhling 2004; Eitel et al. 2005).

New Zealand has the other major loess deposits of the Southern Hemisphere. They cover extensive areas, especially in eastern South Island and southern North Island. It has been estimated that loess more than 1 m thick covers at least 10% of New Zealand's land surface and that soils with a loessial component cover 60% of the country (Eden and Hammond 2003). The loess has been derived mainly from dust deflated by westerly winds from the many broad, braided river floodplains. Dust is deflated from point bars and abandoned channels and deposited downwind on the floodplains. Some of the loess may have been derived from the continental shelf at times of low glacial sea levels. New Zealand loess has a predominantly quartzo-feldspathic mineralogy and is largely derived from uplifted Mesozoic turbidite sequences from the main axial ranges and uplifted Neogene marine sequences, though in the North Island particularly the loess also contains a tephra (volcanic ash) component. Some of the New Zealand loess is of considerable antiquity, and in the Wanganui region of North Island there is a 500×10^3 year record of 11 loess layers and associated palaeosols (Palmer and Pillans 1996). On South Island, luminescence studies suggest that the Romahapa loess/palaeosol sequence is at least 350×10^3 years old (Berger et al. 2002). However, dust continues to accumulate in New Zealand at the present time downwind of many major braided floodplains; and the maximum thickness of post-glacial loess on the Canterbury Plains is about 4 m (Berger et al. 1996).

In Europe, the loess is most extensive in the east where, as in the case of North America, there were plains and steppe conditions. The German loess shows a very close association with outwash and, in France, the same situation is observed along the Rhône and Garonne Rivers. These two rivers carried outwash from glaciers in the Alps and Pyrenees, respectively. The Danube was another major source of silt for loess in eastern Europe. Britain has relatively little loess and this may have resulted from the oceanic climate which would tend to reduce the area of exposed outwash. Indeed, in Britain wind-lain sediments of periglacial times are conspicuous only for their rarity and "loess is more of a contaminant of other deposits than one in its own right" (Williams 1975). The maximum depth of loess in Britain is only about 2–3 m. In southern Europe, Late Pleistocene loess, up to 10 m thick, occurs in the Granada Basin of south-east Spain (Günster et al. 2001). Other loess is known from the central Apennines of Italy (Frezzotti and Giraudi 1990), the Po Valley (Busacca and Cremaschi 1998; Castiglioni 2001), Susak Island in the Dalmatian Archipelago (Cremaschi 1990) and in parts of Greece, including Crete (Brunnacker 1980).

Loess is probably more widespread in South Asia than has often been realised. Given the size of the Thar Desert and the large amounts of sediment that are transported to huge alluvial plains by rivers draining from the mountains of High Asia, this is scarcely surprising. In northern Pakistan, there are loess deposits in the Potwar Plateau (Rendell 1989) and in Kashmir there are many loess – palaeosol sequences (Dilli and Pant 1994) while, in north India, loess has been identified from the Delhi Ridge of Rajasthan (Jayant et al. 1999), various tributary valleys of the Ganges plain, such as the Son and the Belan (Williams and Clarke 1995) and the central Himalayas (Pant et al. 2005). It has also been found in the plains of Gujarat in western India (Malik et al. 1999).

We will now first consider the controversial matter of loess and its relative paucity on the margins of the world's greatest contemporary dust source and then will look at the huge loess deposits of Central Asia and of China.

10.2   PeriSaharan Loess

Although loess (by definition a wind-deposited dust with a median grain size range of 20–30 μm; Tsoar and Pye 1987) has been estimated to cover up to 10% of the world's land area (Pesci 1968), its occurrence in Africa is very limited. This appears surprising, given that the Sahara is the world's largest area of contemporary dust storm activity; and evidence from ocean and ice cores suggests that it produced more dust during the cold phases of the Pleistocene.

The reasons for the relative lack of loess deposits around the Sahara are a subject for debate (see Wright 2001b). Some have argued that sufficient silt-sized material could only be produced in glacial environments and that the Sahara lacks loess because it has few mountains and therefore receives insufficient material from mountain glaciers (Smalley and Krinsley 1978). This is unlikely to be the full explanation because, as we saw in Section 2.2, there are many mechanisms whereby silt is produced in deserts and there is self-evidently plenty of silt in the Sahara at the present day to provide material for dust storm transport (McTainsh 1987; Tsoar and Pye 1987; Yaalon 1987). Certainly much Saharan dust has been deposited over the oceans (Fig. 10.2), but on land only certain desert margins appear to have been favourable for loess formation. Tsoar and Pye (1987) suggest that globally the absence of more widespread peridesert loess is largely due to a lack of available vegetation traps for dust, an idea also put forward by Coudé-Gaussen (1990) in comparing loess deposits north and south of the Mediterranean. Another possible reason is the relative high intensity of rainfall (and therefore of water erosion) on the south side of the Sahara. The mean rainfall per rainy day in the drier parts of West Africa averages 9.75 mm, whereas in the drier parts (mean annual rainfall less than 400 mm) of the classic loess belts it is 4.51 mm (China) and 2.56 mm (former USSR).

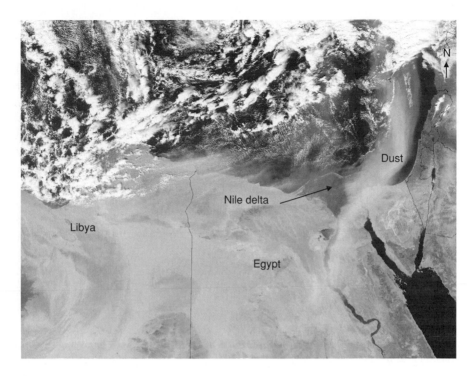

Fig. 10.2. A dust storm blowing northeastwards into the Mediterranean from North Africa, 2 February 2003 (MODIS). Much Saharan dust has been deposited over the oceans but this is not a complete explanation for the relative lack of PeriSaharan loess

Several authors suggest that the current inventory of loess derived from the Sahara is incomplete (e.g. Coudé-Gaussen 1987; Yaalon 1987), but three areas have been studied in some detail: southern Tunisia (Coudé-Gaussen et al. 1982), Northern Nigeria (McTainsh 1987) and the Negev (Yaalon and Dan 1974). The Matmata plateau loess (Fig. 10.3) of southern Tunisia reaches a thickness of 18 m at Téchine and contains up to five palaeosols typically rich in smectite and palygorskite. The loess probably derives from the Sabkha, Chott Djerid and from the Grand Erg Oriental.

Coudé-Gaussen et al. (1983) suggested that two great phases of deposition occurred between 28 000 years BP and 10 000 years BP and from 6000 years BP to 4000 years BP; and Coudé-Gaussen (1991) provides full details of their sedimentology. However, while Coudé-Gaussen et al. (1983) believed that maximum loess deposition occurred during humid conditions, this view was disputed by Dearing et al. (1996) on the basis of their mineral magnetics investigation. They believed that the period between 15 000 years BP and 20000 years BP was a time of both aridity and accelerated loess deposition. More recently, Dearing et al. (2001) showed that some of the loess is older

Fig. 10.3. The loess deposits of the Matmata area in Tunisia have been excavated to create dwellings (from ASG)

than this, with a sequence of loess and palaeosols from Téchine being deposited during the period between 100 000 years BP and 250 000 years BP.

The silty loess of the Jebel Gharbi mountain range in north-west Libya, a deposit that reaches a maximum thickness of 4–5 m and contains interbedded palaeosols and calcretes (Giraudi 2005), is effectively an extension of the Matmata loess. Elsewhere in Libya, a clayey loess has been documented in the Ghat area in the south-west (Assallay et al. 1996).

On the south side of the Sahara, material from the Chad basin transported by the Harmattan wind system has provided the source of the Zaria loess mantle of the Kano plain in northern Nigeria, which displays a clear decrease in grain size with distance from the basin. The dominant clay minerals in the Zaria loess are illite and kaolinite (McTainsh 1987).

Other sparse deposits are catalogued by Coudé-Gaussen (1987): (a) to the north of the Sahara in the Canary Islands, Southern Morocco, south-western Egypt and (b) to the south in Guinea and Northern Cameroon. In the Negev Desert of the Middle East, the Netivot loess section is up to 12 m thick and contains distinct palaeosols of Upper Pleistocene and Holocene age, which indicate climatic cycles of about 20 000 years duration. Here the dominant clay mineral is montmorillonite, with some pedogenic palygorskite. Loess has also been identified in the central Sinai (Rögner and Smykatz-Klosss 1991). Some of these Near Eastern dust deposits have an origin that is at least in part African.

10.3   Central Asian Loess

One of the most striking features of Central Asia, and one it shares with China (Bronger et al. 1998), is the development of very thick (some more than 200 m thick) and complex loess deposits dating back to the Pliocene (Ding et al. 2002; Fig. 10.4). They are well displayed in both the Tajik Republic (Mestdagh et al. 1999) and the Uzbek Republic (Zhou et al. 1995), where rates of deposition were very high in late Pleistocene times (Lazarenko 1984). The nature of the soils and pollen grains preserved in the loess profiles suggest a progressive trend towards greater aridity through the Quaternary; and this may be related to progressive uplift of the Ghissar and Tien Shan mountains (see Davis et al. 1980). A thermoluminescence (TL) chronology for the Middle and Upper Pleistocene loess deposits of Tajikistan is provided by Frechen and Dodonov (1998) and section and granulometric details are provided by Goudie et al. (1984). However, some of the early TL dates for the deposits are believed to be unreliable (Dodonov and Baiguzina 1995; Zhou et al. 1995). None the less, as in China, the loess profiles contain a large number of palaeosols that formed during periods of relatively moist and warm climate. Rates of loess deposition were very modest in the Holocene whereas, in the Last Glacial, rates of accumulation were as high as 1.20 m per 1000 years (Frechen and Dodonov 1998). Ding et al. (2002) believe that the alternations

Fig. 10.4. The loess deposits of Khonako, Tajikistan (from ASG)

of loess and soil horizons in Central Asia can be well correlated with the Chinese loess and deep-sea isotope records.

10.4 Chinese Loess

Loess (*huangtu*, yellow earth) reaches its supreme development in China, most notably in the Loess Plateau (Fig. 10.5), a 450 000 km^2 area in the middle reaches of the Yellow River (Hwang Ho). At Jiuzhoutai, north-west of Lanzhou, the loess attains a maximum thickness of 334 m, while in Jingyuan County, Gansu Province, a thickness of 505 m has been reported (Huang et al. 2000), but over most of the plateau 150 m is more typical. The loess, because of its mechanical properties, creates distinctive landscapes, but it is also important because it provides one of the best terrestrial records of past climates. The classic study is that of Liu (1988). Loess deposits occur in locations other than the Loess Plateau, including the mountainous regions (Rost 1997; Lehmkuhl 1997; Sun 2002b), the Tibetan Plateau (Lehmkuhl et al. 2000),

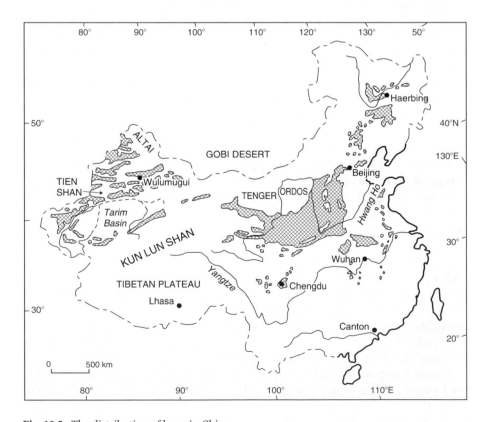

Fig. 10.5. The distribution of loess in China

parts of northern Mongolia (Feng 2001) and Korea (Yatagai et al. 2002). The loess of China poses many challenges for the engineer because of the development of pseudo-karst, landslides and huge sediment yields in stream channels (Derbyshire and Meng 2005).

In some areas, loess *sensu stricto* overlies the Pliocene Red Clay Formation (PRCF) which is also in part a product of aeolian dust accumulation (Liu et al. 2003; Yang and Ding 2004). Evidence for this is that the 'red clay' has similar particle size characteristics to the palaeosols that occur within the overlying loess deposits. Its base has been dated to around $7.2–8.35\times10^6$ years ago (Qiang et al. 2001). It covers an area of 400 000 km^2 and ranges in thickness from 10 m to more than 100 m (Lu et al. 2001). Although the clay was thought to mark the start of aeolian dust accumulation in China and the onset of the present-day East Asian monsoon system (Sun et al. 1998; An 2000; Ding and Yang 2000), it seems that Chinese deserts and their production of dust actually date back much further. Dust derived from the Tibetan Plateau and the Gobi is evident in ocean core deposits going back to at least 11×10^6 years BP (Pettke et al. 2000), while aeolian deposits in Qinan County in Gansu Province indicate that deserts large enough to produce significant dust output must have been formed by 22×10^6 years ago in central Asia (Guo et al. 2002).

The boundary between the loess and the PRCF has been palaeomagnetically dated at 2.5×10^6 years ago. The abrupt commencement of loess deposition on a large scale at about 2.5×10^6 years ago implies a major change in atmospheric conditions and the ongoing uplift of the Tibetan Plateau may have contributed to this (Ding et al. 1992). The appearance of loess beds alternating with numerous palaeosols indicates a cyclical climatic regime, with dry cold conditions being dominated by the north-westerly monsoon and humid warm conditions being dominated by the south-easterly monsoon. This contrasts with the more continuous warm climate that prevailed in the preceding 3×10^6 years during the Pliocene. The Nd and Sr isotopic composition of the aeolian deposits changed at around 2.58×10^6 years ago; and this has been attributed by Sun (2005) to the addition of relatively younger crustal materials to the dust in response to the climatic cooling and late Cainozic uplift, which promoted glacial grinding in the high orogenic belts of central Asia.

It appears that the accumulation of aeolian dust accelerated rapidly from about 1.2×10^6 years ago and that the front of loess deposition was pushed 600 km further south-eastwards from 0.6×10^6 years ago (Huang et al. 2000). At the Jiaxian section (Qiang et al. 2001), rates of sedimentation were about 6 m per million years between 5.0×10^6 years ago and 3.5×10^6 years ago, rising to 16 m per million years between 3.5×10^6 years ago and 2.58×10^6 years ago and reaching 20–30 m per million years thereafter.

Immediately above the PRCF is the Wucheng Loess. Above that in turn are the Lower Lishi Loess, the Upper Lishi Loess and the youngest unit, the Malan Loess (late Pleistocene). There may also have been some relatively limited Holocene loess deposition, but average rates of loess accumulation in the

Loess Plateau were higher, possibly by a factor of two, in the later part of the last glacial period than during the Holocene (Pye and Zhou 1989). The last glacial appears to have been a time when soil moisture contents were low, dunes became destabilised and the desert margin shifted southwards towards the Loess Plateau (Rokosh et al. 2003).

The loess units contain large numbers of palaeosols with as many as 32 soils present above the PRCF (Fig. 10.6). Differences in the nature of these soils and of the loess in between have been used to establish the history of climate over the last 2.5×10^6 years (Liu and Ding 1998). The loess can furnish a high resolution record of change so that sub-millennial-scale variations have been picked up (Heslop et al. 1999). Porter (2001) has argued that high-frequency fluctuations in dust influx during the period of Malan dust deposition may be correlated with North Atlantic Heinrich events. At longer time-scales, various periodicities have been identified in Chinese loess – palaeosol sequences, associated with orbital fluctuations, including 100×10^3-year and 400×10^3-year cycles (Lu et al. 2004).

Figure 10.7 indicates the relationship between loess and palaeosol sequences, loess magnetic susceptibility and the oxygen isotope record from the Pacific Ocean. In general terms, periods of loess deposition are associated with cold phases (which by implication are dry), while the palaeosols are associated with warmer phases (An et al. 1990; Sartori et al. 2005), indicating their origin as products of deflation and subsequent transport and deposition by dust storms. During the last glacial cycle, it was westerly and north-westerly winds that were the most important agents for the transport of dust to the Loess Plateau (Lu and Sun 2000). A comparison of the magnetic signatures of the loess with sands from the Taklamakan suggests that some of the loess was derived from that source region (Torii et al. 2001), while the presence of calcareous nanofossils in the Malan Loess suggests transport by westerly winds from the Tarim basin (Zhong et al. 2003).

In addition to palaeosols, the Loess Plateau sections show multiple phases of gully formation and gully infilling; and these have been interpreted by Porter and An (2005) in terms of phases of drainage incision under moist, intensified summer-monsoon conditions and phases of gully-infilling by loess during glacial, cold-dry winter-monsoon conditions.

The grain size characteristics of the loess change in a southerly (Yang and Ding 2004) and easterly direction, with the coarsest loess (mean grain size ca. 33 µm) being deposited by north-westerly winds in close proximity to the inner Asian deserts. By contrast, the loess in the south-eastern part of the Loess Plateau has a mean size that is only 15 µm, while the median diameter on Cheju island, Korea, ranges from 6 µm to 16 µm (Yatagai et al. 2002). Likewise, the thickness of the Malan Loess declines progressively along a WNW–ESE transect as one moves away from the desert source regions and into areas with higher levels of precipitation (Porter 2001). Grain size also varies down section and may give information on past wind velocities (Nugteren et al. 2004; Sun et al. 2004). Coarser grains are correlated with cold

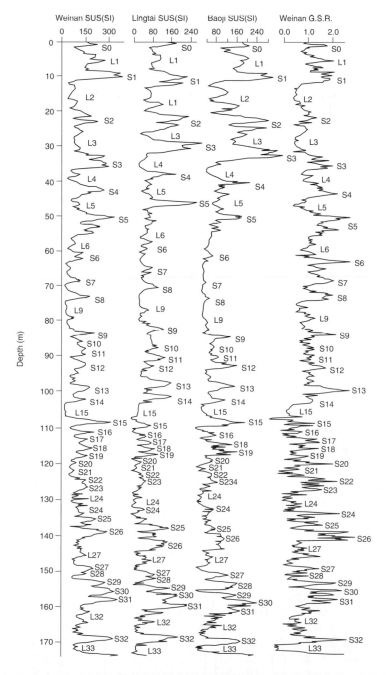

Fig. 10.6. Correlation of magnetic susceptibility curves along Chinese loess sections (*SI*) and the grain-size ratio of the <2 μm fraction to that of the <10 μm fraction. The major units of loess (*L* units 1–33) and soil (*S* units 0–32) are indicated. Modified after Liu and Ding (1998, Figs. 5, 6)

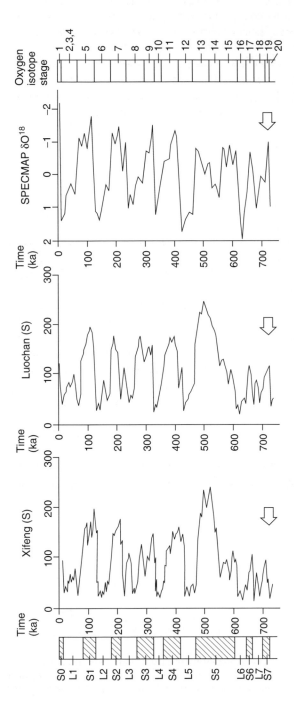

Fig. 10.7. A Comparison between magnetic susceptibility profiles at Xifeng and Luochuan, the stratigraphy of loess (*L*) and soil (*S*) layers at Xifeng, the SPECMAP oxygen isotope record from a North Pacific marine core and the oxygen isotope 'time-scale'. Modified after Pye and Sherwin (1999, Fig. 10.9), copyright John Wiley and Sons Ltd. Reproduced with permission

periods characterised by an increased winter-monsoon strength such as Heinrich Events and the Younger Dryas, whereas finer grains coincide with periods of enhanced summer-monsoon circulation, such as interstadials (Yatagai et al. 2002).

The dust that formed the Chinese loess appears to have been trapped downwind by an *Artemisia*-dominated grassland vegetation through the past 130–170×10^3 years (Jiang and Ding 2005; Zhang et al. 2006) with but sparse evidence that, over the same period, there was a widespread forest cover (Liu et al. 2005). C4 plant abundance declined during glacials, but increased during palaeosol formation in interglacials (Vidic and Montañez 2004).

10.5 Conclusions

The great loess deposits of China and other parts of the world give an indication of the importance that dust storms have played in moulding the Earth's surface. Although we live in a dusty world today, the evidence from loess, ice cores, lake sediments and ocean cores all indicate that dust storm activity has from time to time been very substantially greater than in the contemporary era. Over the past few decades, analysis of climatological data and remote sensing imagery has given us a range of new insights into the nature and distribution of present-day dust activity, so that we can now say a great deal about the distribution of dust storms, their source areas, their trajectories of movement and their frequencies. We are also beginning to learn why it is that dust storm activity varies on decadal timescales in response to climate changes and varying degrees of human influence. Perhaps most importantly of all, we can now appreciate that the dust derived from the world's deserts plays a major role in the Earth System through its contribution to biogeochemical cycling and climate; and we can appreciate the direct role that dust plays in human affairs, including the conduct of warfare and the spread of disease.

References

Abdulaziz AO (1994) A study of three types of wind-blown dust in Kuwait: duststorms, rising dust and suspended dust. J Meteorol 19:19–23

Abouchami W, Zable M (2003) Climate forcing of the Pb isotope record of terrigenous input into the Equatorial Atlantic. Earth Planet Sci Lett 213:221–234

Abrantes F (2003) A 340,000 year continental climate record from tropical Africa – news from opal phytoliths from the equatorial Atlantic. Earth Planet Sci Lett 209:165–179

Acosta-Martínez V, Zobeck TM (2004) Enzyme activities and arylsulfatase protein content of dust and the soil source: biochemical fingerprints? J Environ Qual 33:1653–1661

Acworth RI, Jankowshi J (2001) Salt source for dryland salinity – evidence from an upland catchment in the Southern Tablelands of New South wales. Aust J Soil Sci 39:39–59

Adams JW, Rodriguez D, Cox RA (2005) The uptake of SO_2 on Saharan dust: a flow tube study. Atmos Chem Phys 5:2679–2689

Afeti GM, Resch FJ (2000) Physical characteristics of Saharan dust near the Gulf of Guinea. Atmos Environ 34:1273–1279

Agence France Press (1985) Tchad – divers; fog de sable sur Ndjamena, 16th February. Agence France Press, Ndjamena

Alastuey A, et al. (2005) Characterisation of TSP and PM2.5 at Izaña and Sta Cruz de Tenerife (Canary Islands, Spain) during a Saharan dust episode (July 2002). Atmos Environ 39:4715–4728

Aléon J, Chaussidon M, Marty B, Schütz L, Jaenicke R (2002) Oxygen isotopes in single micrometer-sized quartz grains: tracing the source of Saharan dust over long-distance atmospheric transport. Geochim Cosmochim Acta 66: 3351–3365

Alexander J (1931) The fatal Belgian fog. Science 73:96

Al-Harthi AA, Bankher KA (1999) Collapsing loess-like soil in western Saudi Arabia. J Arid Environ 41:383–399

Al-Najim FA (1975) Dust storms in Iraq. Bull Coll Sci 16:437–451

Alpert P, Ganor E (1993) A jet stream associated heavy dust storm in the eastern Mediterranean. J Geophys Res 98:7339–7349

Alpert P, Ganor E (2001) Sahara mineral dust measurements from TOMS. Comparison to surface observations over the Middle East for the extreme dust storm, March 14–17, 1988. J Geophys Res 106:18275–18286

Alpert P, Kaufman YJ, Shay-El Y, Tanre D, Silva A da, Schubert S, Joseph JH (1998) Quantification of dust-forced heating of the lower troposphere. Nature 395:367–370

Alpert P, Herman J, Kaufman YJ, Carmona I (2000) Response of the climatic temperature to dust forcing, inferred from total ozone mapping spectrometer (TOMS) aerosol index and the NASA assimilation model. Atmos Res 53:3–14

An Z (2000) The history and variability of the East Asian palaeomonsoon climate. Quat Sci Rev 19:171–187

An Z, Liu T, Lu Y, Porter SC, Kukla G, Wu X, Hua Y (1990) The long-term paleomonsoon variation recorded by the Loess–Palaeosol sequence in central China. Quat Int 7/8:91–95

Anderson K, Wells S, Graham R (2002) Pedogenesis of vesicular horizons, Cima volcanic field, Mojave Desert, California. J Soil Sci Soc Am 66:878–887

Anderson R (1987) Grassland revegetation by land imprinting: a new option in desertification control. Desert Control Bull 14:38–44

Andreae MO (1995) Climatic effects of changing atmospheric aerosol levels. In: Henderson-Sellers A (ed) World survey of climatology vol 16. Future climates of the world. Elsevier, Amsterdam, pp. 341–392

Andreae MO (1996) Raising dust in the Greenhouse. Nature 380:389–390

Andrews E, Stone R, Dutton E, Andreson G, Harris J, Shettle E, Ogren J (2003) Asian dust signatures at Barrow: observed and simulated. In: Abstracts of the 2nd workshop on mineral dust. LISA/CNRS, Paris

Anon (1935) Dust storms, April 1935. Mon Weather Rev 63:148

Anon (1977) Neige rouge et pluie de boue sur le Dauphine. J Meteorol 2:211–212

Aoki I, Kurosaki Y, Osada R, Sato T, Kimura F (2005) Dust storms generated by mesoscale cold fronts in the Tarim Basin, Northwest China. Geophys Res Lett 32, L06807, doi: 10.1029/2004GL021776

Arimoto MO (2001) Eolian dust and climate: relationships to sources, tropospheric chemistry, transport and deposition. Earth Sci Rev 54:29–42

Arimoto R, Ray BJ, Lewis NF, Tomza U, Duce RA (1997) Mass particle size distribution of atmospheric dust and the dry deposition of dust to the remote ocean. J Geophys Res 102:15867–15874

Armbrust DV, Dickerson JD (1971) Temporary wind erosion control: cost and effectiveness of 34 commercial materials. J Soil Water Conserv 26:154–157

Armbrust DV, Lyles L (1975) Soil stabilizers to control wind erosion. Soil Sci Soc Am Spec Publ 7:77–82

Asian Development Bank (2005) Establishment of a regional monitoring and early warning network for dust and sandstorms in northeast Asia, volume 2. Asian Development Bank, Manila

Assallay AM, Rogers CDF, Smalley IJ (1996) Engineering properties of loess in Libya. J Arid Environ 32:373–386

Aston SR, Chester R, Johnson LR, Padgham RC (1973) Eolian dust from the lower atmosphere of the eastern Atlantic and Indian Oceans, China Sea and Sea of Japan. Mar Geol 14:15–28

Avila A, Peñuelas J (1999) Increasing frequency of Saharan rains over northeastern Spain and its ecological consequences. Sci Total Environ 228:153–156

Avila A, Rodà F (2002) Assessing decadal changes in rainwater alkalinity at a rural Mediterranean site in the Montseny Mountains (NE Spain). Atmos Environ 36:2881–2890

Avila A, Queralt I, Gallart F, Martin-Vide J (1996) African dust over northeastern Spain: mineralogy and source regions. In: Guerzoni S, Chester R (eds) The impact of desert dust across the Mediterranean. Kluwer, Dordrecht, pp. 201–205

Avila A, Queralt-Mitjans I, Alarcón M (1997) Mineralogical composition of African dust delivered by red rains over northeastern Spain. J Geophys Res 102: 21977–21996

Aymoz G, Jaffrezo JL, Jacob V, Colomb A, George C (2004) Evolution of organic and inorganic components of aerosol during a Saharan dust episode observed in the French Alps. Atmos Chem Phys 4:2499–2512

Babaev AG (1996) Problems of arid land development. Moscow University Press, Moscow

Bach AJ, Brazel AJ, Lancaster N (1996) Temporal and spatial aspects of blowing dust in the Mojave and Colorado deserts of southern California, 1973–1994. Phys Geogr 17:329–353

Bagnold RA (1933) A further journey through the Libyan Desert. Geogr J 82:103–126

Bagnold RA (1941) The physics of blown sand and desert dunes. Methuen, London

Bain DC, Tait NM (1977) The mineralogy and origin of dust-fall on Skye. Clay Miner 12:353–355

Balakirev EK (1968) Unusual dust storms over Ashkhabadom (in Russian). Priroda 10:124–155

Baldwin CS (1988) The influence of field windbreaks on vegetable and specialty crops. Agric Ecosyst Environ 22/23:191–203

Balme M, Metzger S, Towner M (2003a) Friction wind speeds in dust devils. Geophys Res Lett 30:1830

Balme MR, Whelley PL, Greeley R (2003b) Mars: dust devil track survey in Argyre Planitia and Hellas Basin. J Geophys Res 108:E8, doi: 10.1029/2003JE002096

Banoub EF (1970) Sandstorms and duststorms in UAR. Technical Note 1. United Arab Republic Meteorological Department, Cairo, 35 pp

Banzon VF, Evans RE, Gordon HR, Chomko RM (2004) SeaWifs observations of the Arabian Sea southwest monsoon bloom for the year 2000. Deep-sea research part II. Top Stud Oceanogr 51:1189–1208

Barkan J, Kutiel H, Alpert P (2004) Climatology of dust sources in North Africa and the Arabian Peninsula, based on TOMS data. Indoor Built Environ 13:407–419

Barkan J, Alpert P, Kutiel H, Kishcha P (2005) Synoptics of dust transportation days from Africa towards Italy and central Europe. Journal of Geophysical Research 110, D07208, doi: 10.1029/2004JD005222

Bärring L, Jönsson P, Mattsson JO, Åhman R (2003) Wind erosion on arable land in Scania, Sweden and the relation to the wind climate – a review. Catena 52:173–190

Bar-Ziv J, Goldberg GM (1974) Simple siliceous pneumoconiosis in Negev Bedouins. Arch Environ Health 29:121–126

Becker J (1986) Gobi N-tests blamed for high cancer toll, 28th January. The Guardian, London, p. 11

Belnap J, Gillette DA (1997) Disturbance of biological soil crusts: impacts on potential erodibility of sandy desert soils in southeastern Utah. Land Degrad Dev 8:355–362

Belnap J, Gillette DA (1998) Vulnerability of desert biological soil crusts to wind erosion: the influence of crust development, soil texture and disturbance. J Arid Environ 39:133–142

Bennett HH (1938a) Soil conservation. McGraw Hill, New York

Bennett HH (1938b) Emergency and permanent control of wind erosion in the Great Plains. Sci Mon 47:381–399

Bergametti G, Gomes L, Remoudaki E, Desbois M, Martin D, Buat-Ménard P (1989) Present transport and deposition patterns of African dusts to the north-western

Mediterranean. In: Leinen M and Sarnthein M (eds) Palaeoclimatology and palaeometeorology: modern and past patterns of global atmospheric transport. NATO ASI Ser C 282:227-252

Berger GW, Tonkin PJ, Pillans BJ (1996) Thermoluminescence ages of post-glacial loess, Rakaia River, South Island, New Zealand. Quat Int 34/36:177-181

Berger GW, Pillans BJ, Bruce JG, McIntosh PD (2002) Luminescence chronology of loess-paleosol sequences from southern South island, New Zealand. Quat Sci Rev 21:1899-1913

Bertainchand E (1901) Sur les poussières atmosphériques observées a Tunis le Mars 1901. C R Acad Sci 132:1153-1155

Bertrand J, Cerf A, Domergue JL (1979) Repartition in space and time of dust haze south of the Sahara. WMO Bull 538:409-415

Bettis EA, Muhs DR, Roberts HM, Wintle AG (2003) Last glacial loess in the conterminous USA. Quat Sci Rev 22:1907-1946

Betzer PR, Carder KL, Duce RA, Merrill JT, Tindale NW, Uematsu M, Costello DK, Young RW, Feely RJ, Breland JA, Bernstein RE, Greco AM (1988) Long-range transport of giant mineral aerosol-particles. Nature 336:568-571

Bielders CL, Michels K, Rajot J-R (2000) On-farm evaluation of ridging and residue management practices to reduce wind erosion in Niger. J Soil Sci Soc Am 64:1776-1785

Bielders CL, Lamers JPA, Michels K (2001) Wind erosion control technologies in the West African Sahel: the effectiveness of windbreaks, mulching and soil tillage, and the perspective of farmers. Ann Arid Zone 40:369-394

Biscaye PE, Chesselet R, Prospero JM (1974) Rb-Sr, (87)Sr/(86)Sr isotope system as an index of provenance of continental dusts in the open Atlantic Ocean. J Rech Atmos 8:819-824

Bishop JKB, David RE, Sherman JT (2002) Robotic observations of dust storm enhancement of carbon biomass in the North Pacific. Science 298:817-820

Blanco A, De Tomasi F, Filippo E, Manno D, Perrone MR, Serra A, Tafuro AM, Tepore A (2003) Characterization of African dust over southern Italy. Atmos Chem Phys 3:2147-2159

Blank RR, Young JA, Allen FL (1999) Aeolian dust in a saline playa environment, Nevada, USA. J Arid Environ 41:365-381

Bloemendal J, King JW, Hunt A, DeMenocal PB, Hayashida A (1993) Origin of the sedimentary magnetic record at ocean drilling program sites on the Owen Ridge, western Arabian Sea. J Geophys Res 98:4199-4219

Blumberg DG, Greeley R (1996) A comparison of general circulation model predictions to sand drift and dune orientations. J Climate 9:3248-3259

Blümel WD (1982) Calcretes in Namibia and SE Spain: relations to substratum, soil formation and geomorphic factors. Catena Suppl 1:67-82

Blümel WD (1991) Kalkkrusten - ihre genetischen Beziehungen zu Bodenbildungen und äolischer Sedimentation. Geomethodica 16:169-197

Bonasoni P, Cristofanelli P, Calzolari F, Bonaf'e U, Evangelisti F, Stohl A, Dingenen R van, Colombo T, Balkanski Y (2004) Aerosol-ozone correlations during dust transport episodes. Atmos Chem Phys Discuss 4:2055-2088

Bonatti E, Arrhenius G (1965) Eolian sedimentation in the Pacific off northern Mexico. Mar Geol 3:337-348

Bonnifield P, (1979) The Dust Bowl. University of New Mexico Press, Albuquerque

Boon KF, Kiefert L, McTainsh GH (1998) Organic matter content of rural dusts in Australia. Atmos Environ 32:2817-2823

Bopp L, Kohfeld KE, Le Quere C (2003) Dust impact on marine biota and atmospheric CO2 during glacial periods. Palaeoceanography 18:1046

Borbély-Kiss I, Kiss AZ, Koltay E, Szabó G, Bozó L (2004) Saharan dust episodes in Hungarian aerosol: elemental signatures and transport trajectories. J Aerosol Sci 35:1205–1224

Borushko IS (1972) Dust storm distribution in the tropics (in Russian). Glav Geofiz Obs (Leningrad) 284:76–83

Bory AJ-M, Biscaye PE, Svensson A, Grousset FE (2002) Seasonal variability in the origin of recent atmospheric mineral dust at North GRIP, Greenland. Earth Planet Sci Lett 196:123–134

Bowden LW, Huning JR, Hutchinson CF, Johnson CW (1974) Satellite photograph presents first comprehensive view of local wind: the Santa Ana. Science 184:1077–1078

Boyé M, Marmier F, Nesson C, Trécolle G (1978) Les depôts de la Sebkha Mellala. Rev Geomorphol Dyn 27:49–62

Bozzano G, Kuhlmann H, Alonso B (2002) Storminess control over African dust input to the Moroccan Atlantic margin (NW Africa) at the time of maxima boreal summer insolation: a record of the last 220 kyr. Palaeogeogr Palaeoclimatol Palaeoecol 183:155–168

Brandtner E (1947) Der Staubfall in Westeuropa am 29 Marz, 1947. Meteorole Rundsch 1:222–226

Brazel AI (1991) Blowing dust and highways: the case of Arizona, USA. In: Perry AH, Symons J (eds) Highway meteorology. Spon, London, pp. 131–161

Brazel A, Hsu S (1981) The climatology of hazardous Arizona dust storms. In: Péwé TL (ed) 1981 Desert dust. Geol Soc Am Spec Pap 186:293–303

Breuning-Madsen H, Awadzi TW (2005) Harmattan dust deposition and particle size in Ghana. Catena 63:23–38

Briat M, Royer A, Petit JR, Lorius C (1982) Late glacial input of eolian continental dust in the Dome C ice core: additional evidence from individual microparticle evidence. Ann Glaciol 3:27–31

Brimhall GH, Lewis CJ, Ague JJ, Dietrich W.E, Hampel J, Teague T, Rix P (1988) Metal enrichment in bauxites by deposition of chemically mature aeolian dust. Nature 333:819–824

Brimhall GH, Lewis CJ, Ford C, Brad J, Taylor G, Warin O (1991) Quantitative geochemical approach to pedogenesis: importance of parent material reduction, volumetric expansion, and eolian influence in laterization. Geoderma 51:51–91

Broecker WS (2002) Dust: climate's Rosetta Stone. Proc Am Philos Soc 146:77–80

Bronger A, Winter R, Heinkele T (1998) Pleistocene climatic history of East and Central Asia based on paleopedological indicators in loess-palaeosol sequences. Catena 34:1–17

Brookes IA (1993) Geomorphology and Quaternary geology of the Dakhla oasis region, Egypt. Quat Sci Rev 12:529–552

Brooks CEP (1920) The meteorology of British Somaliland. Q J R Meteorol Soc 46:434–438

Brooks N (1999) Dust–climate interactions in the Sahel–Sahara zone of northern Africa, with particular reference to late twentieth century Sahelian drought. PhD thesis, University of East Anglia, Norwich

Brooks N, Legrand M (2000) Dust variability over northern Africa and rainfall in the Sahel. In: McLaren SJ, Kniverton D (eds) Linking land surface change to climate change. Kluwer, Dordrecht, pp. 1–25

Brown JKM, Hovmøller MS (2002) Aerial dispersal of pathogens on the global and continental scales and its impact on plant disease. Science 297:537–541

Brown MJ, Krauss RK, Smith RM (1968). Dust deposition and weather. Weatherwise 21:66–69

Brunnacker K (1980) Young Pleistocene loess as an indicator for the climate in the Mediterranean area. Palaeoecol Africa 12:99–113

Bryant R (2003) Monitoring hydrological controls on dust emissions: preliminary observations from Etosha Pan, Namibia. Geogr J 169:131–141

Bryson RA, Barreis DA (1967) Possibilities of major climatic modifications and their implications: northwest India, a case for study. Bull Am Meteorol Soc 48:136–142

Buch MW, Zoller L (1992) Pedostratigraphy and thermoluminescence-chronology of the western margin-(Lunette-)dunes of Etosha Pan, Northern Namibia. Wurzburger Geogr Arb 84:361–384

Bücher A (1986) Recherches sur les poussières minerals d'origine saharienne. PhD thesis, Université de Reims-Champagne-Ardenne, Reims

Bücher A, Lucas G (1975) Poussières Africaines sur l'Europe. Meteorologie 33:53–69

Bücher A, Lucas G (1984) Sédimentation éolienne intercontinentale, poussières sahariennes et géologie. Bull Cent Rech Explor Prod Elf Aquitaine 8:151–165

Bücher A, Dessens J (1992) Saharan dust over France and England, 6–9 March 1991. J Meteorol 17:226–233

Bullard JE, White K (2005) Dust production and the release of iron oxides resulting from the Aeolian abrasion of natural dune sands. Earth Surf Process Landforms 30:95–106

Bullard JE, McTainsh GH, Pudmenky C (2004) Aeolian abrasion and modes of fine particle production from natural red dune sands: an experimental study. Sedimentology 51:1103–1125

Bullock MS, Larnet FJ, Cesar Izaurralde R, Feng Y (2001) Overwinter changes in wind erodibility of clayloam soils in southern Alberta. J Soil Sci Soc Am 65:423–430

Burritt B, Hyers AD 1981 Evaluation of Arizona's dust warning system. In: Péwé TL (ed) Desert dust. Geol Soc Am Spec Pap 186:281–292

Burt S (1991a) Falls of dust rain in Berkshire, March 1991. Weather 46:248

Burt S (1991b) Falls of dust rain within the British Isles. Weather 46:347–353

Burt S (1992) Fall of dust rain, 11 October 1991. Weather 47:142

Burt S (1995) Fall of dust rain in Berkshire, 24 September 1994. Weather 50:30–31

Busacca A, Cremaschi M (1998) The role of time versus climate in the formation of the deep soils of the Apennine fringe of the Po valley, Italy. Quat Int 51/52: 95–107

Cahill TA (1984) Studies of particulate matter at Mono Lake. Final report to the California Air Quality Resources Board on contract A1-144-32. California Air Quality Resources Board, David, 16 pp

Cakmur RV, Miller RL, Tegen I (2001) A comparison of seasonal and interannual variability of soil dust aerosols over the Atlantic Ocean as inferred by the TOMS AI and AVHRR AOT retrievals. J Geophys Res 106:18287–18303

Callot Y, Marticorena B, Bergametti G (2000) Geomorphologic approach for modelling the surface features of arid environments in a model of dust emissions: application to the Sahara desert. Geodin Acta 13:245–270

Calvo E, Pelejero C, Logan GA, De Deckker P (2004) Dust-induced changes in phytoplankton composition in the Tasman Sea during the last four glacial cycles. Paleoceanography 19:PA2020

Cannon WF, Dean WE, Bullock JH (2003) Efects of Holocene climate change on mercury deposition in Elk Lake, Minnesota: the importance of eolian transport in the mercury cycle. Geology 31:187–190

Cantor BA (2003) MGS-MOC observations of Martian dust storm activity. Int Conf Mars 6:3166

Capo RC, Chadwick OA (1999) Sources of strontium and calcium in desert soil and calcrete. Earth Planet Sci Lett 170:61–72

Caquineau S, Gaudichet A, Gomes L, MagonthierM-C, Chatenet B (1998) Saharan dust:clay ratio as a relevant tracer to assess the origin of soil-derived aerosols. Geophys Res Lett 25:983–986

Caquineau S, Gaudichet A, Gomes L, Legrand M (2002) Mineralogy of Saharan dust transported over northwestern tropical Atlantic Ocean in relation to source regions. J Geophys Res 107, doi: 10.1029/2000JD000247

Carlson TN (1979) Atmospheric turbidity in Saharan dust outbreaks as determined by analyses of satellite brightness data. Mon Weather Rev 107:322–335

Carlson TN, Prospero JM (1972) The large-scale movement of Saharan air outbreaks over the northern equatorial Atlantic. J Appl Meteorol 11:283–297

Carrico CM, Bergin MH, Shrestha AB, Dibb JE, Gomes L, Harris JM (2003) The importance of carbon and mineral dust to seasonal aerosol properties in the Nepal Himalaya. Atmos Environ 37:2811–2824

Castiglioni GB (2001) Forme e depositi di origine eolica. Suppl Geogr Fis Din Quat 4:119–122

Cattle SR, McTainsh GH, Wagner S (2002) Aeolian dust contributions to soil of the Namoi Valley, northern NSW, Australia. Catena 47:245–264

Cecil CB (2004) Eolian dust and the origin of sedimentary chert. US Geol Surv Open File Rep 2004–1098, 11 pp

Cehak K (1976) Ein staubbriederschlag in Mitteleuropa. Wetter Leben 28:905–907

Chamley H (1988) Contribution éolienne à la sedimentation marine au large du Sahara. Bull Soc Geol Fr 6:1091–1100

Champollion M (1965) Rétombées de poussières et pluies colorées. Meteorologie 70:307–313

Chan YC, McTainsh G, Leys J, McGowan H, Tews K (2005) Influence of the 23 October 2002 dust storm on the air quality of four Australian cities. Water Air Soil Pollut 164:329–348

Changery MJ (1983) A dust climatology of the Western United States. NOAA, Ashville, N.C., 25 pp

Changnon SA (1983) Record dust storms in Illinois: causes and implications. J Soil Water Conserv 38:58–63

Chazette PL, Pelon J, Moulin C (2001) Lidar and satellite retrieval of dust aerosols over the Azores during SOFIA/ASTEX. Atmos Environ 35:4297–4304

Chen W, Fryrear DW (2002) Sedimentary characteristics of a haboob dust storm. Atmos Res 61:75–85

Chen W, Fryrear DW, Yang Z (1999) Dust fall in the Taklamakan Desert of China. Phys Geogr 20:189–224

Chen XY, Boaler JM, Magee JW (1991) Gypsum ground: a new occurrence of gypsum sediment in playas of central Australia. Sediment Geol 72:79–95

Chen YS, Sheen PC, Chen ER, Liu YK, Wu TN, Yang CY (2004) Effects of Asian dust storm events on daily mortality in Taipeh, Taiwan. Environ Res 95:151–155

Chepil WS (1945) Dynamics of wind erosion. Soil Sci 60:305–320, 397–411, 475–480

Chepil WS, Milne RA (1941) Wind erosion of soil in relation to roughness of surface. Soil Sci 52:417–431

Chepil WS, Woodruff NP (1963) The physics of wind erosion and its control. Adv Agron 15:211–302

Chepil WS, Siddoway FH, Armbrust DV (1962) Climate factor for estimating wind erodibility of farm fields. J Soil Water Conserv 17:162–165

Chester R, Johnson LR (1971) Atmospheric dust collected off the Atlantic coasts of North Africa and the Iberian Peninsula. Mar Geol 11:251–260

Chester R, Elderfield H, Griffin JJ (1971) Dust transported in the north-east and south-east trade winds in the Atlantic Ocean. Nature 233:474–476

Chester R, Elderfield JJ, Griffin JJ, Johnson LR, Padgham RC (1972) Eolian dust along the eastern margins of the Atlantic Ocean. Mar Geol 13:91–106

Chester R, Sharples EJ, Sanders GS, Saydam AC (1984) Saharan dust incursion over the Tyrrhenian Sea. Atmos Environ 18:929–935

Chester R, Berry AS, Murphy KJT (1991) The distributions of particulate atmospheric trace metals and mineral aerosols over the Indian Ocean. Mar Chem 34:261–290

Chiapello I, Moulin C (2002) TOMS and METEOSAT satellite records of the variability of Saharan dust transport over the Atlantic during the last two decades (1979–1997). Geophys Res Lett 29:1176

Chiapello I, Prospero JM, Herman JR, Hsu NC (1999) Detection of mineral dust over the North Atlantic Ocean and Africa with the Nimbus 7 TOMS. J Geophys Res 104:9277–9291

Chiapello I, Moulin C, Prospero JM (2005) Understanding the long-term variability of African dust transport across the Atlantic as recorded in both Barbados surface concentrations and large-scle total ozone mapping spectrometer (TOMS) optical thickness. J Geophys Res 110, doi:10.1029/2004JD005132

Chin M, Chu A, Levy R, Remer L, Kaufman Y, Holben B, Eck T, Ginoux P, Gao Q (2004) Aerosol distribution in the Northern hemisphere using ACE-Asia: results from global model, satellite observations, and sun photometer measurements. J Geophys Res 109:D23S90, doi:10.1029/2004JD004829

Chiquet A, Fabrice C, Hamelin B, Michard A, Nahon N (2000) Chemical mass balance of calcrete genesis on the Toledo granite (Spain). Chem Geol 170:19–35

Chizhikov, YA, Kamlyuk, GG (1997) Peculiarities of the dust storms occurrence in Belarus. Lithosphere 6:80–92

Chlachula J (2003) The Siberian loess record and its significance for reconstruction of Pleistocene climate change in north-central Asia. Quat Sci Rev 22:1879–1906

Choi H, Choi DS (2005) Prediction of duststorm evolution by vorticity theory. Proc SympAir Qual Manage Urban Reg Global Scales 3:26–38

Choun HF (1936) Dust storms in southwestern plains area. Mon Weather Rev 64:195–199

Christopher SA, Wang J, Xia X (2004) Ground-based observations of dust optical properties in the Chinese dust source region and intercomparison with MISR aerosol retrievals. Proc Conf Satellite Meteorol Oceanogr 13:1–5

Chun Y (2003) Historical records of Asian dust events in Korea. In: Abstracts of the 2nd workshop on mineral dust. LISA/CNRS, Paris

Chun Y, Cho KM (2003) Asian dust events in Korea for recent 100 years. In: Abstracts of the 2nd workshop on mineral dust. LISA/CNRS, Paris

Chun Y, Boo K-O, KimY, Park S-U, Lee M (2001) Synopsis, transport and physical characteristics of Asian dust in Korea. J Geophys Res 106:18461–18469

Chung YS, Kim HS, Park KH, Jhun JG, Chen SJ (2003) Atmospheric loadings, concentrations and visibility associated with sandstorm: satellite and meteorological analysis. Water Air Soil Pollut Focus 3:21–40

Cinderey MB (1987) Dust deposits in England on 9 November 1984. Weather 42:28

Clafin LE, Stuteville DL, Armburst DV (1973) Windblown soil in the epidemiology of bacterial leaf spot of alfalfa and common blight of beans. Phytopathology 63:1417–1419

Claquin T, Roelandt C, Kohlfeld KE (2003) Radiative forcing by ice-age atmospheric dust. Clim Dyn 20:183–202

Clark I, Assamoi P, Bertrand J, Giorgi F (2004) Characterization of potential zones of dust generation at eleven stations in the southern Sahara. Theor Appl Climatol 77:173–184

Clarke FW (1916) The data of geochemistry, 3rd edn. United States Geological Survey, Washington, D.C.

Claustre H, et al. (2002) Is desert dust making oligotrophic waters greener? Geophys Res Lett 29, doi: 10.1029/2001GL014056

Clayton JD, Clapperton CM (1997) Broad synchrony of a late-glacial advance and the highstand of palaeolake Tauca in the Bolivian Altiplano. J Quat Sci 12:169–182

Clemens SC, Prel WL (1990) Late Pleistocene variability of Arabian Sea summer monsoon winds and continental aridity: eolian records from the lithogenic component of deep sea sediments. Palaeoceanography 5:109–145

Clements T, Stone RO, Mann JF, Eymann JL (1963) A study of windborne sand and dust in desert areas. Technical Report ES-8. US Army Natick Laboratories, Earth Science Division, Natick, Mass.

Coakley JA, Cess RD (1985) Response of the NCAR community climate model to the radiative forcing by the naturally occurring tropospheric aerosol. J Atmos Sci 42:1677–1692

Coffey M (1978) The dust storms. Nat Hist (NY) 87:72–83

Colarco PR, Toon OB, Torres O, Rasch PJ (2002) Determining the UV imaginary index of Saharan dust particles from TOMS data and a three dimensional model of dust transport. J Geophys Res 107:10.1029

Collaud Coen M, Weingartner E, Schaub D, Hueglin C, Corrigan C, Henning S, Schwikowski M, Baltensperger U (2004) Saharan dust events at the Jungfraujoch: detection by wavelength dependence of the single scattering albedo and first climatology analysis. Atmos Chem Phys 4:2465–2480

Collyer FX, Barnes BG, Churchman GJ, Clarkson TS, Steiner JT (1984) Trans-Tasman dust transport event. Weather Clim 4:42–46

Conca JL, Rossman GR (1982) Case hardening of sandstone. Geology 10:520–523

Conte M, Colacino M, Piervitali E (1996) Atlantic disturbances deeply penetrating the African continent: effects over Saharan regions and the Mediterranean Basin. In: Guerzoni S, Chester R (eds) The impact of desert dust across the Mediterranean. Kluwer, Dordrecht, pp. 93–102

Cooke RU, Brunsden D, Doornkamp JC (1982) Geomorphological hazards in urban drylands. Clarendon Press, Oxford

Cooke RU, Warren A, Goudie A (1993) Desert geomorphology. UCL Press, London

Coque R (1962) La Tunisie présahrienne, étude géomorphologique. Armand Colin, Paris

Coque-Delhuille BC, Gentelle PP (1998) Aeolian dust and superficial formations in the arid part of Yemen. In: Alsharhan AR, Glennie K, Whittle G, Kendall CG (eds) Quaternary deserts and climatic change. Balkema, Rotterdam, pp. 199–208

Cornelis WM, Gabriels D (2005) Optimal windbreak design for wind-erosion control. J Arid Environ 61:315–332

Coudé-Gaussen G (1981) Etude détaillée d'un échantillon de poussières éoliennes prélevé au Tanezrouft, le 10 Décembre 1980. Rech Geogr Strasbourg 16/17:121–130

Coudé-Gaussen G (1984) Le cycle des poussières éoliennes désertiques actuelles et la sedimentation des loess péridesertiques quaternaries. Bull Cent Rech Explor Prod Elf Aquitaine 8:167–182

Coudé-Gaussen G (1987) The periSaharan loess: sedimentological characterisation and palaeoclimatic significance. GeoJournal 15:177–183

Coudé-Gaussen G (1989) Local, proximal and distal Saharan dusts: characterization and contribution to the sedimentation. In: Leinen M, Sarnthein M (eds) Paleoclimatology and paleometeorology: modern and past patterns of global atmospheric transport (NATO ASI C282). NATO, Washington, D.C., pp. 339–358

Coudé-Gaussen G (1990) The loess and loess-like deposits along the sides of the western Mediterranean Sea: genetic and palaeoclimatic significance. Quat Int 5:1–8

Coudé-Gaussen G (1991) Les poussières Sahariennes. John Libbey Eurotext, Montrouge

Coude-Gaussen G, Blanc P (1985) Présence de grains éolisés et palygorskite dans les poussières actuelles et les sédiments recents d'origine désertique. Bull Soc Geol Fr 1:571–579

Coudé-Gaussen G, Rognon P (1988) Origine éolienne de certains encroûtements calcaires sur l'ile de Fuerteventura (Canaries Orientales). Geoderma 42:271–293

Coudé-Gaussen G, Mosser C, Rognon P, Torenq J (1982) Une accumlation de loess Pleistocene supérieur dans le sud-Tunisien: coupe de Téchine. Bull Soc Geol Fr 24:283–292

Coudé-Gaussen G, Olive L, Rognon P (1983) Datation de dépôts loessiques et varations climatiques à la bordure nord du Sahara algéro-tunisien. Rev Geomorphol Dyn Geogr Phys 24:61–73

Coudé-Gaussen G, Désiré E, Regrain R (1988) Particularité des poussières sahariennes distales tombées sur la Picardie et l'Ile de France le 7 mai 1988. Hommes Terres Nord 4:246–251

Cremaschi M (1990) Stratigraphy and palaeenvironmental significance of the loess deposits on Susak Island (Dalmatian Archipelago). Quat Int 5:97–105

Criado C, Dorta P (2003) An unusual 'blood rain' over the Canary Islands (Spain). The storm of January 1999. J Arid Environ 55:765–783

Crooks GA, Cowan GRC (1993) Duststorm, South Australia, November 7th, 1988. Bull Aust Meteorol Oceanogr Soc 6:68–72

Cullen HM, deMenocal PB, Hemkming S, Hemming G, Brown FH, Guilderson T, Sirocko F (2000) Climate change and the collapse of the Akkadian empire: evidence from the deep sea. Geology 28:379–382

Dahl KA, Oppo DW, Eglington TI, Hughen KA, Curry WB, Sirocko F (2005) Terrigenous plant wax inputs to the Arabian Sea: implications for the reconstruction of winds associated with the Indian Monsoon. Geochim Cosmochim 69:2547–2558

D'Almeida GA (1986) A model for Saharan dust transport. J Climate Appl Meteorol 25:903–916

D'Almeida GA (1987) Desert aerosol characteristics and effects on climate. In: Leinen M, Sarnthein M (eds) Palaeoclimatology and palaeometeorology: modern and past patterns of global atmospheric transport. NATO ASI Ser C 282:311–338

Darby DA, Burckle LH, Clark DL (1974) Airborne dust on the Arctic pack ice, its composition and fallout rate. Earth Planet Sci Lett 24:166–172

Dare-Edwards AJ (1984) Aeolian clay deposits of south-eastern Australia: parna or loessic clay? Trans Inst Br Geogr 9:337–344

Daremova K, Sokolik IN, Darmenov A (2005) Characteristics of east Asian dust outbreaks in the spring of 2001 using ground-based and satellite data. J Geophys Res 110, D02204, doi: 10.1029/2004JD004842

Darwin C (1846) An account of the fine dust which often falls on vessels in the Atlantic Ocean. Q J Geol Soc 2:26–30

Darwin C (1889) Journal of Researches (5th edn). Ward Lock, London

Darwin C (1893) Journal of Researches. T. Nelson and Sons, London

Das TM (1988) Effects of deposition of dust particles on leaves of crop plants on screening of solar illumination and associated physiological processes. Environ Pollut 53:421–422

Dave JV (1978) Effect of aerosols on the estimation of total ozone in an atmospheric column from the measurement of its ultraviolet radiation. J Atmos Sci 35: 899–911

Davey P (2004) A short note on the great immigration of February 2004. http://www.dorsetbutterflies.co.uk/paintedlady_migration_2004.htm

Davis RS, Ranov VA, Dodonov AE (1980) Early man in Soviet Central Asia. Sci Am 243:91–102

Davitaya FF (1969) Atmospheric dust content as a factor affecting glaciation and climate change. Ann Assoc Am Geogr 59:552–560

Dayan U (1986) Climatology of back trajectories from Israel based on synoptic analysis. J Climate Appl Meteorol 25:591–595

Dayan U, Heffter J, Miller J, Gutman G (1991) Dust intrusion into the Mediterranean basin. J Appl Meteorol 30:1185–1199

De Angelis M, Gaudichet A (1991) Saharan dust deposition over Mont Blanc (French Alps) during the last 30 years. Tellus 43B:61–75

Dearing JA, Livingstone IP, Zhou LP (1996) A late Quaternary magnetic record of Tunisian loess and its climatic significance. Geophys Res Lett 23:189–192

Dearing JA, Livingstone IP, Bateman MD, White K (2001) Palaeoclimate records from OIS 8.0–5.4 recorded in loess in palaeosol sequences on the Matmata Plateau, southern Tunisia, based on mineral magnetism and new luminescence dating. Quat Int 76/77:43–56

De Graaf M, Stammes P, Torres O, Koelemeijer RBA (2005) Absorbing aerosol index: sensitivity analysis, application to GOME and comparison with TOMS. J Geophys Res 110, doi:10.1029/2004JD005178

Delany AC, Parkin DW, Goldberg ED, Riemann BEF, Griffin JJ (1967) Airborne dust collected at Barbados. Geochim Cosmochim Acta 31:895–909

Delgado R, Martin-Garcia JM, Oyonarte C, Delgado G (2003) Genesis of the *terrae rossae* of the Sierra Gádor (Andalusia, Spain). Eur J Soil Sci 54:1–16

Delmas V, Jones HG, Tranter M, Delmas R (2005) The weathering of aeolian dusts in alpine snows. Atmosph Environ 30:1317–1325

Delmonte B, Basile-Doelsch I, Petit J-R, Maggi V, Revel-Rolland M, Michard A, Jagoutz E, Grousset F (2004a) Comparing the Epica and Vostok dust records during the last 220,000 years: stratigraphical correlation and provenance in glacial periods. Earth Sci Rev 66:63–87

Delmonte B, Petit JR, Andresen KK, Basile-Doelsch I, Maggi V, Ya Lipenkov V (2004b) Dust size evidence for opposite regional atmospheric circulation changes over East Antarctica during the last climatic transition. Climate Dyn 23:427–438

DeMenocal PB (1995) Plio-Pleistocene African climate. Science 270:53–59

DeMenocal P, Ortiz J, Guilderson T, Sarnthein M (2000) Coherent high- and low-latitude climate variability during the Holocene warm period. Science 288:2198–2202

Derbyshire E (2001) Geological hazards in loess terrain, with particular reference to the loess regions of China. Earth Sci Rev 54:231–260

Derbyshire E, Meng X (2005) Loess. In: Fookes PG, Lee EM, Milligan G (eds) Geomorphology for engineers. CRC, Boca Raton, pp. 688–728

Derbyshire E, Meng X, Kemp RA (1998) Provenance, transport and characteristics of modern aeolian dust in western Gansu Province, China, and interpretation of the Quaternary loess record. J Arid Environ 39:497–516

De Rouw A, Rajot J-L (2004) Soil organic matter, surface crusting and erosion in Sahelian farming systems based on manuring or fallowing. Agric Ecosyst Environ 104:263–276

Dessens J, Van Dinh P (1990) Frequent Saharan dust outbreaks north of the Pyrenees: a sign of climatic change? Weather 45:327–333

De Wet LW (1979) Die invloed van die Suid-Afrikaanse hoogland op die ontwikkeling en voortplanting van atmosferiese druk en sirkulasiestelsels. MSc thesis, University of Pretoria, Pretoria

Dey S, Tripathi SN, Singh RP, Holben BN (2004) Influence of dust storms on the aerosol optical properties over the Indo-Gangetic basin. J Geophys Res 109, doi: 10.1029/2004JD004924

Dickerson JD, Woodruff NP, Banbury EE (1976) Techniques for improving survival and growth of trees in semiarid areas. J Soil Water Conserv 31:63–66

Dilli K, Pant RK (1994) Clay minerals as indicators of the provenance and palaeoclimatic record of the Kashmir loess. J Geol Soc India 44:563–574

Ding RQ, Li JP, Wang SG, Ren FM (2005) Decadal change of the spring dust storm in northwest China and the associated atmospheric circulation. Geophys Res Lett 32:L02808

Ding Z, Yang SL (2000) C3/C4 vegetation evolution over the last 7.0 Myr in the Chinese Loess Plateau: evidence from pedogenic carbonate $\delta13C$. Palaeogeogr Palaeoclimatol Palaeoecol 160:292–299

Ding Z, Rutter N, Han J, Liu T (1992) A coupled environmental system formed at about 2.5 Ma in East Asia. Palaeogeogr Palaeoclimatol Palaeoecol 94:223–242

Ding ZL, Ravov V, Yang SL, Finaev A, Han JM, Wang GA (2002) The loess record in southern Tajikistan and correlation with Chinese loess. Earth Planet Sci Lett 200:387–400

Dodonov AE, Baiguzina LL (1995) Loess stratigraphy of Central Asia: palaeoclimatic and palaeoenvironmental aspects. Quat Sci Rev 14:707–720

Donarummo J, Ram M, Stoermer EF (2003) Possible deposit of soil dust from the 1930s USA dust bowl identified in Greenland ice. Geophys Res Lett 30, doi:10.1029/2002GL016641

Doornkamp JC, Brunsden D, Jones DKC (1980) Geology, geomorphology, and pedology of Bahrain. GeoAbstracts, Norwich

Dorn R (1986) Rock varnish as an indicator of aeolian environmental change. In: Nickling WG (ed) Aeolian geomorphology. Allen and Unwin, Winchester, Mass., pp. 291–307

Drab E, Gaudichet A, Jaffrezo JL, Colin JL (2002) Mineral particles content in recent snow at Summit (Greenland). Atmos Environ 36:5365–5376

Drees LR, Manu A, Wilding LP (1993) Characteristics of aeolian dusts in Niger, West Africa. Geoderma 59:213–233

Dubief J (1953) La vent et la deplacement du sable du Sahara. Trav Inst Rech Sahariennes 8:123–164

Dubief J (1953) Les vents de sable dans le Sahara Français. Coll Int CNRS 35:45–70

Duce RA (1995) Sources, distributions, and fluxes of mineral aerosols and their relationship to climate. In: Charlson RJ, Heintzenberg J (eds) Aerosol forcing of climate. Wiley, Chichester, pp. 43–72

Duce RA, Unni CK, RayEJ, Prospero JM, Merrill JT (1980) Long range transport of soil dust from Asia to the tropical North Pacific: temporal variability. Science 209:1522–1524

Duffour A (1931) Sur la pluie du 27 au 28 novembre 1930, dans la région pyrénéenne. Bull Soc Hist Nat Toulouse 61

Dunion JP, Velden CS (2004) The impact of the Saharan Air Layer on Atlantic tropical cyclone activity. Bull Am Meteorol Soc 85:353–365

Durkee PA, Pfeil F, Frost E, Shema R (1991) Global analysis of aerosol particle characteristics. Atmosph Environ 25A:2557–2471

Durst CS (1935) Dust in the atmosphere. Q J R Meteorol Soc 61:81–89

Du Toit AL (1906) Geological survey of portions of the divisions of Vryburg and Mafeking. Annu Rep Geol Comm Cape Good Hope 10:205–258

Eckardt F (1996) The distribution and origin of gypsum in the central Namib Desert, Namibia. D Phil thesis, University of Oxford, Oxford

Eckardt F, Drake N, GoudieAS, White K, Viles H (2001) The role of playas in pedogenic gypsum crust formation in the central Namib Desert: a theoretical model. Earth Surf Process Landforms 26:1177–1193

Eckardt F, Washington R, Wilkinson MJ (2002) The origin of dust on the west coast of southern Africa. Palaeoecol Afr Surrounding Islands 27:207–219

Eden DN, Hammond AP (2003) Dust accumulation in the New Zealand region since the last glacial maximum. Quat Sci Rev 22:2037–2052

Ehrenberg CG (1849) Passat Staub und Blutregen. Abhandlungen Akademie Wissenschaften, Berlin

Eitel B, Blumel W, Huser K (2001) Dust and loessic alluvial deposits in northwestern Namibia (Damaraland, Kaokoveld): sedimentology and paleoclimatic evidence based on luminescence data. Quat Int 76/77:57–65

Eitel B, Hecht S, Mächtle B, Schukraft G, Kadereit A, Wagner G A, Kromer B, Unkel I, Reindel M (2005) Geoarchaeological evidence from desert loess in the Nazca-Palpa region, southern Peru: palaeoenvironmental changes and their impact on pre-Columbian cultures. Archaeometry 47:1–137

Ekström M, McTainsh GH, Chappell A (2004) Australian dust storms: temporal trends and relationships with synoptic pressure distributions (1960-99). Int J Climatol 24:1581–1599

El-Askary HM, Sarkar S, Kafatos M, El-Ghazawi TA (2003) A multisensor approach to dust storm monitoring over the Nile Delta. IEEE Trans Geosci Remote Sensing 41:2386–2391

El-Askary H, Gautam R, Singh RP, Kafatos M (2005) Dust storms detection over the Indo-Gangetic basin using multi sensor data. Adv Space Res (in press)

Embabi NS (1999) Playas of the Western desert, Egypt. Ann Acad Fenn Geol Geogr 160:5–47

Engelstädter S (2001) Dust storm frequencies and their relationship to land surface conditions. Diploma thesis, Friedrich-Schiller-University, Jena

EPICA community members (2004) Eight glacial cycles from an Antarctic ice core. Nature 429:623–628

Eppink LAAJ (1982) A survey of wind and water erosion in the Netherlands and an inventory of Dutch erosion research. (Rep Dept Land Water Use 59) Agricultural University, Wageningen

Ervin RJ, Lee JA (1994) Impact of conservation practices on airborne dust in the southern High Plains of Texas. J Soil Water Conserv 49:430–437

Escudero M, Castillo S, Querol X, Avila M, Alarcón M, Viana MM, Alastuey A, Cuevas E, Rodríguez S (2005) Wet and dry African dust episodes over eastern Spain. J Geophys Res 110, doi:10.1029/2004JD004731

Fahey B (1985) Salt weathering as a mechanism of rock breakup in cold climates: an experimental approach. Zeitschr Geomorphol 29:99–111

Fan K, Wang HJ (2004) Antarctic Oscillation and the dust weather frequency in north China. Geophys Res Lett 31:L10201

Fan S-M, Horowitz LW, Levy H, Moxim WJ (2004) Impact of air pollution on wet deposition of mineral dust aerosols. Geophys Res Lett 31:L02104

Fang M, Zheng M, Wang F, Chim KS, Kot SC (1999) The long-range transport of aerosols from northern China to Hong Kong – a multi-technique study. Atmos Environ 33:1803–1817

Fei J, Zhou J, Zhang Q, Chen H (2005) Dust weather records in Beijing during 1860–1898 AD based on the diary of Tonghe Weng. Atmos Environ 39:3943–3946

Feng ZD (2001) Gobi dynamics in the Northern Mongolian Plateau during the past 20,000+ yr: preliminary results. Quat Int 76/77:77–83

Ferguson WS, Griffin JJ, Goldberg ED (1970) Atmospheric dusts from the North Pacific – a short note on a long-range eolian transport. J Geophys Res 75:1137–1139

Fett W (1958) Der atmosphärische Staub. Deutscher Verlag der Wissenschaften, Berlin

File RF (1986) Dust deposits in England on 9 November 1984. Weather 41:191–195

Fiol LA, Fornós JJ, Gelabert B, Guijarro JA (2005) Dust rains in Mallorca (Western Mediterranean): their occurrence and role in some recent geological processes. Catena 63:64–84

Fisher WB (1978) The Middle East, 7th edn. Methuen, London

Foda MA, Khalaf FI, Al-Kadi AS (1985) Estimation of dust fallout rates in the northern Arabian Gulf. Sedimentology 32:595–603

Folger DW, Burckle LH, Heezen BC (1967) Opal phytoliths in North Atlantic dust fall. Science 155:1243–1244

Folk RL (1975) Geological urban hindplanning; an example from a Hellenistic Byzantian city, Stobi, Jugoslavian Macedonia. Environ Geol 1:5–22

Forman SL, Oglesby R, Webb RS (2001) Temporal and spatial patterns of Holocene dune activity on the Great Plains of North America: megadroughts and climate links. Global Planet Change 29:1–29

Foucault A, Mélières F (2000) Palaeoclimatic cyclicity in central Mediterranean Pliocene sediments: the mineralogical signal. Palaeogeogr Palaeoclimatol Palaeoecol 158:311–323

Fouquart YB, Bonnell G, Brogniez G, Buriez JC, Smith L, Morcrette JJ, Cerf A (1987). Observations of Saharan aerosols: results of ECLATS field experiment II. Broadband radiative characteristics of the aerosols and vertical radiative flux divergence. J Appl Meteorol 26:38–52

Frank AB, Harris DG, Willis WO (1977) Growth and yields of spring wheat as influenced by shelter and soil water. Agron J 69:903–906

Franzen LG, Hjelmroos M, Kallberg P, Brorstrom-Lunden E, Juntto S, Savolainen A L (1994) The 'yellow snow' episode of northern Fennoscandia, March 1991 – a case study of long-distance transport of soil, pollen and stable organic compounds. Atmos Environ 28:3587–3604

Franzen LG, Hjelmroos M, Kallberg P, Rapp A, Mattsson JO, Brorström-Lundén E (1995) The Saharan dust episode of south and central Europe, and northern Scandinavia, March 1991. Weather 50:313–318

Frechen M, Dodonov AE (1998) Loess chronology of the Middle and Upper Pleistocene in Tajikistan. Geol Rundsch 87:2–20

Frechen M, Oches EA, Kohfeld, KE (2003) Loess in Europe – mass accumulation rates during the Last Glacial Period. Quat Sci Rev 22:1835–1857

Freeman MH (1952) Duststorms of the Anglo-Egyptian Sudan. Meteorol Off Meteorol Rep 11, 22 pp

Frezzotti M, Giraudi C (1990) Late glacial and halocene aeolian deposits and geomorphological features near Roccaraso (Abruzzo, central Italy). Quat Int 5:89–95

Frumkin A, Stein M (2004) The Sahara–East Mediterranean dust and climate connection revealed by strontium and uranium isotopes in a Jerusalem speleothem. Earth Planet Sci Lett 217:451–464

Fuller J, Marsden L (2004) Practical dust control agent and application for alkaline ponds and playas. SME Annu Meet Preprint 2004:557–590

Gabriel A (1938) The southern Lut and Iranian Baluchistan. Geogr J 92:193–208

Gabriel AP, Martínez-Ordaz VA, Velasco-Rodreguez VM, Lazo-Sáenz JG, Cicero, R (1999) Prevalence of skin reactivity to coccidioidin and associated risks factors in subjects living in a northern city of Mexico. Arch Med Res 30:388392

Gagosian RB, Peltzer ET, Zafirou OC (1981) Atmospheric transport of continentally derived lipids to the tropical North Pacific. Nature 291:312–314

Gaiero DM, Depetris PJ, Probst J-L, Bidart SM, Leleyter L (2004) The signature of river- and wind-borne materials exported from Patagonia to the southern latitudes: a view from REEs and implications for paleoclimatic interpretations. Earth Planet Sci Lett 219:357–376

Galopin R (1951) Les poussières èoliennes tombées à Genève en Avril 1944 et le problème de leur origine. Arch Sci 4:217–226

Ganor E (1994) The frequency of Saharan dust episodes over Tel Aviv, Israel. Atmos Environ 28:2867–2871

Ganor E, Foner HA (1996) The mineralogical and chemical properties and the behaviour of aeolian Saharan dust over Israel. In: Guerzoni S, Chester R (eds) The impact of desert dust across the Mediterranean. Kluwer, Dordrecht, pp. 163–172

Ganor E, Foner HA (2001) Mineral dust concentrations, deposition fluxes and deposition velocities in dust episodes over Israel. J Geophys Res 106/D16:18431–18437

Ganor E, Mamane M (1982) Transport of Saharan dust across the eastern Mediterranean. Atmos Environ 16:581–587

Ganor E, Foner HA, Brenner S, Neeman E, Lavi N (1991) The chemical composition of aerosols settling in Israel following dust storms. Atmos Environ 25:2665–2670

Gao T, Su LJ, Ma QX, Li HY, Li X, Yu X (2003) Climatic analyses on increasing dust storm frequency in the springs of 2000 and 2001 in Inner Mongolia. Int J Climatol 23:1743–1755

Gao Y, Arimoto R, Duce RA, Zhang XY, Zhang GY, An ZS, Chen LQ, Zhou MY, Gu, DY (1997) Temporal and spatial distribution of dust and its deposition in the China Sea. Tellus 49B:172–189

Gao Y, Fan S-M, Sarmiento JL (2003) Aeolian iron input to the ocean through precipitation scavenging: a modelling perspective and its implication for natural iron fertilization in the ocean. J Geophys Res 108, doi:10.1029/2002JD002420

Garrison VH, et al. (2003) African and Asian dust: from desert soils to coral reefs. BioScience 53:469–480

Gatz DF, Prospero JM (1996) A large silicon-aluminium aerosol plume in central Illinois: North African desert dust? Atmos Environ 30:3789–3799

Georg DJ (1981) Dustfall and instability rain over Northern Ireland on the night of 28–29 January 1981. Weather 36:216–217

Gerson R, Amit R (1987) Rates and modes of dust accretion and deposition in an arid zone – the Negev, Israel. In: Frostick L, Reid I (eds) Desert sediments ancient and modern. Blackwell Scientific, Oxford, pp. 157–169

Ghinassi M, Magi M, Sagri M, Singer BS (2004) Arid climate 2.5 Ma in the Plio–Pleistocene Valdarno Basin (Northern Apennines, Italy). Palaeogeogr Palaeoclimatol Palaeoecol 207:37–57

Ghobrial SI (2003) Effect of sand storms on microwave propagation. Nat Telecommun Conf Rec 2:43.5.1–43.5.4

Gilbert A (1992) The Imperial War Museum book of the desert war 1940–1942. Book Club Associates, London

Gilbert GK (1895) Lake basins created by wind erosion. J Geol 3:47–49

Giles J (2005) The dustiest place on Earth. Nature 434:816–819

Gill TE (1996) Eolian sediments generated by anthropogenic disturbances of playas: human impacts of the geomorphic system and geomorphic impacts on the human system. Geomorphology 17:207–228

Gill TE, Cahill TA (1992) Playa-generated dust storms from Owens Lake. In: Hall CA, Doyle-Jones V, Widawski B (eds) The history of water: eastern Sierra Nevada, Owens valley, White–Inyo mountains. University of California, Los Angeles, pp. 63–73

Gill TE, Reynolds RL, Zobeck TM (2000) Measurements of current and historic settled dusts in West Texas. Air Waste Manage Pub VIP 97, 0175, 15 pp

Gillette DA (1979) Environmental factors affecting dust emission by wind. In: Morales C (ed) Saharan dust. Wiley, Chichester, pp. 71–91

Gillette DA (1981) Production of dust that may be carried great distances. In: Péwé TL (ed) Desert dust. Geol Soc Am Spec Pap 186:11–26

Gillette DA, Hanson KJ (1989) Spatial and temporal variability of dust production caused by wind erosion in the United States. J Geophys Res 94/D2:2197–2206

Gillette DA, Adams J, Endo A, Smith D (1980) Threshold velocities for input of soil particles into the air by desert soils. J Geophys Res 85:5621–5630

Gillies JA, Nickling WG, McTainsh GH (1996) Dust concentrations and particle-size characteristics of an intense dust haze event: inland delta region, Mali, West Africa. Atmos Environ 30:1081–1090

Gingele FX, De Deckke P (2005) Late Quaternary fluctuations of palaeoproductivity in the Murray Canyons area, South Australian continental margin. Palaeogeogr, Palaeoclimatol, Palaeoecol 220:361–373

Ginoux P, Torres O (2003) Empirical TOMS index for dust aerosol: applications to model validation and source characterisation. J Geophys Res 108/D17:4534

Ginoux PM, Chin I, Tegen I, Prospero J, Holben M, Dubovik O, Lin S-J (2001) Global simulation of dust in the troposphere: model description and assessment. J Geophys Res 106:20255–20273

Ginoux P, Prospero JM, Torres O, Chin M (2004) Long-term simulation of global dust distribution with the GOCART model: correlation with North Atlantic Oscillation. Environ Model Software 19:113–128

Giraudi C (2005) Eolian sand in peridesert northwestern Libya and implications for Late Pleistocene and Holocene Sahara expansions. Palaeogeogr Palaeoclimatol Palaeoecol 218:161–173

Gisladottir FO, Arnalds O, Gisladottir G (2005) The effect of landscape and retreating glaciers on wind erosion in South Iceland. Land Degrad Dev 16:177–187

Glaccum RA, Prospero JM (1980) Saharan aerosols over the tropical north Atlantic: mineralogy. Mar Geol 37:295–321

Glaisby CP (1971) The influence of aeolian transport of dust particles on marine sedimentation in the south-west Pacific. J R Soc NZ 1:285–300

Glawion H (1939) Staub und Staubfälle in Arosa. Beitr Phys Frei Atmos 25:1–43

Glazovsky NF (1995) Aral Sea. In: Mandych AF (ed) Enclosed seas and large lakes of eastern Europe and Middle Asia. SPB Academic, Amsterdam, pp. 119–154

Glennie KW Singhvi AK (2002) Event stratigraphy, palaeoenvironment and chronology of SE Arabian deserts. Quat Sci Rev 21:853–869

Gobbi GP, Barnaba F, Blumthaler M, Labow G, Herman JR (2002) Observed effects of particles nonsphericity on the retrieval of marine and desert dust aerosol optical depth by lidar. Atmos Res 61:1–14

Godon NA, Todhunter PE (1998) A climatology of airborne dust from the Red River Valley of North Dakota. Atmos Environ 32:1587–1594

Goede A, McCulloch M, McDermott F, Hawkesworth C (1998) Aeolian contribution to strontium and strontium isotope variations in a Tasmanian speleothem. Chem Geol 149:37–50

Goldberg ED, Griffin JJ (1970) The sediments of the northern Indian Ocean. Deep Sea Res 17:513–537

Goodman GT, Inskip MJ, Smith S, Parry GDR, Burton MAS (1979) The use of moss-bags in aerosol monitoring. In: Morales C (ed) Saharan dust. Wiley, Chichester, pp. 211–232

Goossens D (1995) Field experiments of aeolian dust accumulation on rock fragment substrata. Sedimentology 42:391–402

Goossens D (2003) On-site and off-site effects of wind erosion. In: Warren A (ed) Wind erosion on agricultural land in Europe. European Commission, Luxembourg, pp. 29–38

Goossens D (2005) Quantification of the dry aeolian deposition of dust on horizontal surfaces: an experimental omparison of theory and measurements. Sedimentology 52:859–873

Goossens D, Offer ZY (2005) Long-term accumulation of atmospheric dust in rocky deserts. Z Geomorphol 49:335–352

Goossens D, Van Kerschaever E (1999) Aeolian dust deposition on photovoltaic solar cells: the effects of wind velocity and airborne dust concentration on cell performance. Solar Energy 66:277–289

Gordon JC, Murray MR (1964) A notable case of dust in suspension over Cyprus. Meteorol Mag 93:106–115

Goudie AS (1978) Dust storms and their geomorphological implications. J Arid Environ 1:291–310

Goudie AS (1983) Dust storms in space and time. Progr Phys Geogr 7:502–530

Goudie AS (1990) The landforms of England and Wales. Blackwell, Oxford

Goudie AS, Cooke RU, Doornkamp JC (1979) The formulation of silt from quartz dune sand by salt weathering processes in deserts. J Arid Environ 2:105–112

Goudie AS, Day MJ (1980) Dinsintegration of fan sediments in Death Valley, California, by salt weathering. Phys Geogr 1:126–137

Goudie AS, Middleton NJ (1992) The changing frequency of dust storms through time. Clim Change 20:197–225

Goudie AS, Middleton NJ (2000) Dust storms in South West Asia. Acta Univ Car XXXV:73–83

Goudie AS, Middleton NJ (2001) Saharan dust storms: nature and consequences. Earth Sci Rev 56:179–204

Goudie AS, Viles HA (1995) The nature and pattern of debris liberation by salt weathering: a laboratory study. Earth Surf Process Landforms 20:437–449

Goudie AS, Viles HA (1997) Salt weathering hazards. Wiley, Chichester

Goudie AS, Wells GL (1995) The nature, distribution and formation of pans in arid zones. Earth Sci Rev 38:1–69

Goudie AS, Rendell H, Bull PA (1984) The loess of Tajik SSR. In: Miller KJ (ed) Proceedings of the International Karakoram Project. Cambridge University Press, Cambridge, pp. 399–412

Goudie AS, Stokes S, Livingstone I, Bailiff IK, Allison RJ (1993) Post-depositional modification of the linear sand ridges of the West Kimberley area of north-west Australia. Geogr J 159:306–317

GoudieAS, Stokes S, Cook J, Samieh S, El-Rashidi OA (1999) Yardang landforms from Kharga Oasis, south-western Egypt. Zeitschr Geomorphol Suppl 116:97–112

Goudie AS, Parker AG, Bull PA, White K, Al-Farraj A (2000) Desert loess in Ras Al Khaimah, United Arab Emirates. J Arid Environ 46:123–135

Grayson DK (1993) The desert's past. A natural history of the Great Basin. Smithsonian Institution Press, Washington, D.C.

Greeley R, Iversen JD (1985) Wind as a geological process on Earth, Mars, Venus, and Titan. Cambridge University Press, Cambridge

Greeley R, Lancaster N, Lee S, Thomas P (1992) Martian aeolian processes, sediments, and features. In: Kieffer H, Jakosky B, Snyder C, Matthews M (eds) Mars. Univesity of Arizona, Tucson, pp. 730–766

Griffin DW, Garrison VH, Herman JR, Shinn EA (2001) African desert dust in the Caribbean atmosphere: microbiology and public health. Aerobiologia 17:203–213

Griffin DW, Kellogg CA, Garrison VH, Shinn EA (2002) The global transport of dust. Am Sci 90:228–235

Grigoryev AA, Kondratyev KJ (1981) Atmospheric dust observed from space, part 2. WMO Bull 30:3–9

Grini A, Zender CS, Colarco PR (2002) Saltation sandblasting behaviour during mineral dust aerosol production. Geophys Res Lett 29:1868

Grousset FE, Biscaye PE (2005) Tracing dust sources and transport patterns using Sr, Nd and Pb isotopes. Chem Geol 222:149–167

Grousset FE, Rognon P, Coudé-Gaussen G, Pédemay P (1992) Origins of peri-Saharan dust deposits traced by their Nd and Sr isotopic composition. Palaeogeogr Palaeoclimatol Palaeoecol 93:203–212

Grousset FE, Parra M, Bory A, Martinez P, Bertrand P, Shimmield G, Ellam RM (1998) Saharan wind regimes traced by the Sr-Nd isotopic composition of subtropical Atlantic sediments: last glacial maximum vs today. Quat Sci Rev 17:395–409

Grousset FE, Ginoux P, Bory A (2003) Case study of a Chinese dust plume reaching the French Alps. Geophys Sci Rev 30:1277

Gruber N, Sarmiento JL (1997) Global patterns of marine nitrogen fixation and denitrification. Global Biogeochem Cycles 11:235–266

Guerzoni S, Chester R, Dulac F, Herut B, Loye-Pilot MD, Measures C, Mignon C, Molinari E, Moulin C, Rossini P, Saydam C, Soudine A, Ziveri P (1999) The role of atmospheric deposition in the biogeochemistry of the Mediterranean Sea. Progr Oceanogr 44:147–190

Guieu C, Bozec Y, Blain S, Ridame C, Sarthou G, Leblond N (2002) Chemical characterization of the Saharan dust end-member: some biogeochemical implications for the western Mediterranean Sea. Geophys Res Lett, doi:10.1029/2001GLO14454

Güllü G, Doğan G, Tuncel G (2005) Source regions of dust transported to the eastern Mediterranean. Proc Int Symp Air Qual Manage Urban Reg Global Scales 3:59–67

Günster N, Eck P, Skowronek A, Zöller L (2001) Late Pleistocene loess and their paleosols in the Granada Basin, southern Spain. Quat Int 76/77:241–245

Guo Z T, Ruddiman W F, Hao Q Z, Wu H B, Qiao Y S, Zhu RX, Peng S Z, Wei J J, Yuan BY, Liu TS (2002) Onset of Asian desertification by 22 Myr ago inferred from loess deposits in China. Nature 416:159–163

Gupta JP, Aggarural RK, Gupta GN, Kaul P (1983) Effect of continual application of farmyard manure and urea on soil properties and production of pearl millet in western Rajisthan. Ind J Agric Sci 53:53–56

Guzzi R, Ballista G, Di Micolatino W (2001) Aerosol maps from GOME data. Atmos Environ 35:5079–5091

Gyan K, Henry W, Lacaille S, Laloo A, Lamsee-Ebanks C, McKay S, Antoine RM, Monteil MA (2005) African dust clouds are associated with increased paediatric asthma accident and emergency admissions on the Caribbean island of Trinidad. Int J Biometeorol 49:371–376

Hagen LJ, Woodruff NO (1973) Air pollution from dust storms in the Great Plains. Atmos Environ 7:323–332

Halimov M, Fezer F (1989) Eight yardang types in central Asia. Zeitschr Geomorphol 33:205–217

Hall DJ, Upton SJ, Marsland GW (1994) Designs for a deposition gauge and a flux gauge for monitoring ambient dust. Atmos Environ 28:2963–2979

Hand IF (1934) The character and magnitude of the dense dust cloud which passed over Washington, D.C., 11 May 1934. Mon Weather Rev 62:156–157

Harper RJ, Gilkes RJ (2004) Aeolian influences on the soils and landforms of the southern Yilgarn Craton of semi-arid southwestern Australia. Geomorphology 59:215–235

Harrison SP, Kohfeld KE, Roelandt C, Claquin T (2001) The role of dust in climate changes today, at the last glacial maximum and in the future. Earth Sci Rev 54:43–80

Hassan FA, Barich B, Mahmoud M, Hemdan MA (2001) Holocene playa deposits of Farafra Oasis, Egypt, and their palaeoclimatic and geoarchaeological significance. Geoarchaeology 16:29–46

Haywood J, Francis P, Osborne S, Glew M, Loeb N, Highwood E, Tanre D, Myrhe G, Formenti P, Hirst E (2003) Radiative properties and direct radiative effect of

Saharan dust measured by the C-130 aircraft during SHADE: 1. Solar spectrum. J Geophys Res 108, doi:10.1029/2002JD002687

Haywood JM, Allan RP, Culverwell I, Slingo T, Milton S, Edwards J, Clerbaux N (2005) Can desert dust explain the outgoing longwave radiation anomaly over the Sahara during July 2003? J Geophys Res 110, doi:10.1029/2004JD005232

Hedin S (1903) Central Asia and Tibet. Scribners, New York

Helgren DM, Prospero JM (1987) Wind velocities associated with dust deflation events in the Western Sahara. J Climate Appl Meteorol 26:1147–1151

Henriksson AS, Sarnthein M, Eglinton G, Poynter J (2000) Dimethylsulfide production variations over the past 200 k.y. in the Equatorial Atlantic: a first estimate. Geology 28:499–502

Henz JF, Woiceshyn PM (1980) Climatological relationships of severe dust storms in the Great Plains to synoptic weather patterns (Jet Propulsion Laboratory Publication). California Institute of Technology, Los Angeles, pp. 79–97

Herman JR, Bhartia PK, Torres O, Hsu C, Seftor C, Celarier E (1997) Global distribution of UV-absorbing aerosols from Nimbus 7/TOMS data. J Geophys Res 102/D14:16911–16922

Hernández JLD, Hernández JMM (1997) Tasas de deposicion de polvo atmsoferico en un area semiarida del entorno mediterraneo occidental. Estud Geol 53:211–220

Herrmann L, Stahr K, Jahn R (1999)The importance of source region identification and their properties for soil-derived dust: the case of Harmattan dust sources for eastern West Africa. Contrib Atmos Phys 72:141–150

Herut B, Krom M (1996) Atmospheric input of nutrients and dust to the SE Mediterranean. In: Guerzoni S, Chester R (eds) The impact of desert dust across the Mediterranean. Kluwer, Dordrecht, pp. 349–358

Herwitz SR, Muhs DR, Prospero JM (1996) Origin of Bermuda's clay-rich Quaternary paleosols and their paleoclimatic significance. J Geophys Res 101/D18: 23389–23400

Hesse PP (1994) The record of continental dust from Australia in Tasman Sea sediments. Quat Sci Rev 13:257–272

Hesse PP, McTainsh GH (1999) Last glacial maximum to early Holocene wind strength in the mid-latitudes of the southern hemisphere from aeolian dust in the Tasman Sea. Quat Res 52:343–349

Hesse PP, McTainsh GH (2003) Australian dust deposits: modern processes and the Quaternary record. Quat Sci Rev 22:2007–2035

Heslop D, Shaw J, Bloemendal J, Chen F, Wang J, Parker E (1999) Sub-millennial scale variations in East Asian monsoon systems recorded by dust deposits from the north-western Chinese loess plateau. Phys Chem Earth A 24:785–792

Hilling D (1969) Saharan iron ore oasis. Geogr Mag 41:908–917

Hirose K, Sugimura Y (1984) Excess Th-228 in the airborne dust – an indicator of continental dust from the East Asian deserts. Earth Planet Sci Lett 70:110–114

Hodges L, Brandle JR (1996) Windbreaks: an important component in a plasticulture system. Hortic Technol 6:177–81

Holben B, Eck TF, Fraser RS (1991) Temporal and spatial variability of aerosol optical depth in the Sahel region in relation to vegetation remote sensing. Int J Remote Sensing 12:1147–1163

Holz C, Stuut J-BW, Henrich R (2004) Terrigenous sedimentation processes along the continental margin off NW Africa: implications from grain-size analysis of seabed sediments. Sedimentology 51:1145–1154

Honda M, Shimuzu H (1998) Geochemical, mineralogical and sedimentological studies on the Taklimakan Desert sands. Sedimentology 45:1125–1143

Hongisto M, Sofiev M (2004) Long-range transport of dust to the Baltic Sea region. Int J Environ Pollut 22:72–86

Hooghiemstra H (1989) Variations of the N.W. African trade wind regime during the past 140,000 years: changes in pollen flux evidenced by marine sediment records. In: Leinen M, Sarnthein M (eds) Palaeoclimatology and palaeometeorology: modern and past patterns of global atmospheric transport. Kluwer, Dordrecht, pp. 733–770

Houseman J (1961) Dust haze at Bahrain. Meteorol Mag 90:50–52

Hrádek M, Ŝvehlik R (1995) Dust storms in Moravia and Silesia. In: Hrádek M (ed) Natural hazards in the Czech Republic (Studia Geographica 98). Czech Academy of Sciences, Institute of Geonics, Brno, pp. 83–111

Hsu NC, Herman JR Torres O, Holben B, Tanre D, Eck TF, Smirnov A, Chatenet B, Lavenu F (1999) Comparisons of the TOMS aerosol index with sun photometer aerosol optical thickness: results and applications. J Geophys Res 104:6269–6279

Hsu NC, Herman JR, Weaver C (2000) Determination of radiative forcing of Saharan dust using combined TOMS and ERBE data. J Geophys Res 105/D:20649–20661

Huang CC, Pang J, Zhao J (2000) Chinese loess and the evolution of the East Asian monsoon. Progr Phys Geogr 24:75–96

Hulme M (1985) Dust production in the Sahel. Nature 318:488

Huntington E (1907) The pulse of Asia. Houghton, Mifflin and Co., Boston

Husar RB, Prospero JM, Stowe LL (1997) Characterization of tropospheric aerosols over the oceans with the NOAA advanced very high resolution radiometer optical thickness operational product. J Geophys Res 102/D14:16889–16909

Husar RB, Tratt DM, Schichtel BA, Falke SR, Li F D, Jaffe DS, Gassó ST, Gill T, Laulainen NS, Lu F, Reheis MC, Chun Y, Westphal D, Holben BN, Geymard C, McKendry I, Kuring N, Feldman GC, McClain C, Frouin RJ, Merrill J, DuBois, D, Vignola F, Murayama T, Nickovic S, Wilson WE, Sassen K, Sugimoto N (2001) Asian dust events of April 1998. J Geophys Res 106/D:18317–18330

Hussain A, Mir H, Afzal M (2005) Analysis of dust storms frequency over Pakistan during 1961–2000. Pak J Meteorol 2:49–68

Ichoku C, Remer LA, Kaufman YJ (2004) Evaluation of the moderate-resolution imaging spectroradiometer (MODIS) retrievals of dust aerosol over the ocean during PRIDE. J Geophys Res 108/D19:8594

Idso SB (1976) Dust storms. Sci Am 235:108–111, 113–114

Iino N, Kinoshita K, Tupper AC, Yano T (2004) Detection of Asian dust aerosols using meteorological satellite data and suspended particulate matter concentrations. Atmos Environ 38:6999–7008

Ikegami M, Okada K, Zaizen Y, Tsutsumi Y, Makino Y, Jensen JB, Gras J L (2004) The composition of aerosol particles in the middle troposphere over the western Pacific Ocean: aircraft observations from Australia to Japan, January 1994. Atmos Environ 38:5945–5956

In H-J, Park S-U (2003) Estimation of dust emission amount for a dust storm event occurred in April 1998 in China. Water Air Soil Pollut 148:201–221

Ing GKT (1972) A dust storm over central China, April 1969. Weather 27:136–145

Inoue K, Naruse T (1987) Physical, chemical and mineralogical characteristics of modern eolian dust in Japan and rate of dust deposition. Soil Sci Plant Nutr 33:327–345

Irino T, Ikehara K, Katayama H, Ono Y, Tada R (2003) East Asian monsoon signals recorded in the Japan Sea sediments. PAGES News 9:13–14

Iriondo MH (1997) Models of deposition of loess and loessoids in the Upper Quaternary of South America. J S Am Earth Sci 10:71–79

Iriondo M (2000) Patagonian dust in Antarctica. Quat Int 68/71:83–86

Iriondo M, Kröhling D (2004) "New" types of loess, not related to glaciation. In: HWK (ed) International workshop from particle size to sediment dynamics. HWK, Delmenhorst, pp. 83–85

Israelevich P., Ganor E, Levin Z, Joseph PJ (2003) Annual variations of physical properties of dust over Israel. J Geophys Res 108/D13:4381

Iwasaka Y, Minoura H, Nagaya K (1983) The transport and spatial scale of Asian dust-storm clouds: a case study of the dust-storm event of April 1979. Tellus 35B:189–196

Jackson ML, Levelt TWM, Syers JK, Rex RW, Clayton RN, Sherman GD, Uehara G (1971) Geomorphological relationships of tropospherically derived quartz in the soils of the Hawaiian Islands. Proc Soil Sci Soc Am 35:515–525

Jackson ML, Gillette DA, Danielsen EG, Blifford IH, Bryson RA, Syers JK (1973) Global dustfall during the Quaternary as related to environments. Soil Sci 116:135–145

Jacobs PM, Mason JA (2005) Impact of Holocene dust aggradation on A horizon characteristics and carbon storage in loess-derived Mollisols of the Great Plains, USA. Geoderma 125:95–106

Jaeger H, Carnuth W, Georgi B (1988) Observations of Saharan dust at a north Alpine mountain station. J Aerosol Sci 19:1235–1238

Jaenicke R (1979) Monitoring and critical review of the estimated source strength of mineral dust from the Sahara. In: Morales C (ed) Saharan dust. Wiley, Chichester, pp. 233–242

Jaenicke R, Schütz L (1978) Comprehensive study of physical and chemical properties of the surface aerosols in the Cape Verde Islands region. J Geophys Res 83:3585–3599

Jaffe DA, Mahura A, Kelley J, Atkins J, Novelli PC Merrill J (1997) Impact of Asian emissions on the remote North Pacific atmosphere: interpretation of CO data from Shemya, Guam, Midway and Mauna Loa. J Geophys Res 102:28627–28636

Jaffe DA, Snow J, Cooper O (2003) The April 2001 Asian dust events: transport and substantial impact on surface particulate matter concentrations across the United States. EOS Trans 2003: November 18

Jauregui E (1960) Las tolvaneras de la Ciudad de Mexico. Ing Hidraul Mex 14:60–66

Jauregui E (1969) Apectos météorologicos de la contaminación del aire en la Ciudad de Mexico. Ing Hidraul Mex 23:17–28

Jauregui E (1973)The urban climate of Mexico City. Erdkunde 36:298–307

Jauregui E (1989)The dust storms of Mexico City. Int J Climatol 9:169–180

Jauregui E, Klaus D (1982) Urban effects on precipitation in a large metropolis in the tropics: the case study of Mexico City (in German). Erdkunde 36:278–286

Jayant K, Tripathi V, Rajamani V (1999) Geochemistry of the loessic sediments on Delhi ridge, eastern Thar desert, Rajasthan: implications for exogenic processes. Chem Geol 155:265–278

Jeong M-J, Li Z, Chu DA, Tsay S-C (2005) Quality and compatiobility analyses of global aerosol products derived from the advanced very high resolution radiometer and moderate resolution imaging spectroradiometer. J Geophys Res 110, doi:10.1029/2004JD004648

Jiang H, Ding Z (2005) Temporal and spatial changes of vegetation cover on the Chinese loess plateau through the last glacial cycle: evidence from spore–pollen records. Rev Palaeobot Palynol 133:23–37

Jickells TD, Dorling S, Deuser WG, Church, TM, Arimoto R, Prospero JM (1998) Air-borne dust fluxes to a deep water sediment trap in the Sargasso Sea. Global Biogeochem Cycles 12:311–320

Jickells TD, et al. (2005) Global iron connections between desert dust, ocean biogeochemistry, and climate. Science 308:67–71

Johnson AM (1976) The climate of Peru, Bolivia, and Ecuador. In: Schwerdfeger W (ed) World survey of climatology 12. Elsevier, Amsterdam, pp. 147–202

Jones C, Mahowald N, Luo C (2003) The role of easterly waves in African dust transport. J Climate 16:3617–3628

Jones C, Mahowald N, Luo C (2004) Observational evidence of African dust intensification of easterly waves. Geophys Res Lett 31:L17208, doi:10.1029/2004GL020107

Jones DKC (2001) Blowing sand and dust hazard, Tabuk, Saudi Arabia. In: Griffiths JS (ed) Land surface evaluation for engineering practice (Engineering Geology Special Publications 18). Geological Society, London, pp. 123–128

Joseph PV (1982) A tentative model of Andhi. Mausam 33:417

Joseph PV, Raipal DK, Deka SN (1980) 'Andhi', the convective dust storm of northwest India. Mausam 31:431–412

Jouzel J, Masson V, Cattani O, Falourd S, Stievenard M, Stenni B, Longinelli A, Johnsen SJ, Steffenssen JP, Petit JR, Schwander J, Souchez R, Barkov NI (2001) A new 27 ky high resolution East Antarctic record. Geophys Res Lett 28:3199–3202

Judson S (1968) Erosion of the land. Am Sci 56:356–374

Jung SJA, Davies GR, Ganssen GM, Kroon D (2004) Stepwise Holocene aridification in NE Africa deduced from dust-borne radiogenic isotope records. Earth Planet Sci Lett 221:27–37

Junge CE (1958) Atmospheric chemistry. Adv Geophys 4:1–108

Junge C (1979) The importance of mineral dust as at atmospheric constituent. In: Morales C (ed) Saharan dust. Wiley, Chichester, pp. 49–60

Kalashnikova OV, Kahn R, Sokolik IN, Li W-H (2005) Ability of multiangle remote sensing observations to identify and distinguish mineral dust types: optical models and retrievals of optically thick plumes. J Geophys Res 110:D18S14, doi:10.1029/2004JD004550

Kalma JD, Speight JG, Wasson RJ (1988) Potential wind erosion in Australia – a continental perspective. J Climatol 8:411–428

Kalnay E, Kanamitsu M, Kistler R, Collins W, Deaven D, Gandin L, IredellM, Saha S, White G, Woollen J, ZhuY, Leetmaa A, Reynolds R, Chelliah M, Ebisuzaki W, Higgins W, Janowiak J, Mo KC, Ropelewski C, Wang CJ, Jenne R, Joseph D (1996) The NCEP/NCAR 40-year Reanalysis Project. Bull Am Meteorol Soc 77:437–471

Kalu AE (1979) The African dust plume: its characteristics and propagation across West Africa in winter. In: Morales C (ed) Saharan dust: mobilisation, transport and deposition. Wiley, Chichester, pp. 95–118

Kang S, Dahe Q, Mayewski PA, Wake CP, Ren J (2001) Climatic and environmental records from the far east Rongbuk ice core, Mt. Qomolangma (Mt. Everest). Episodes 24:176–181

Kar A, Takeuchi K (2004) Yellow dust: an overview of research and felt needs. J Arid Environ 59:167–187

Karyampudi VM, Palm SP, Reagan JA, Fung H, Grant WB, Hoff RM, Moulin C, Pierce HF, Torres O, Browell EV, Melfi SH (1999) Validation of the Saharan dust plume conceptual model using LIDAR, Meteosat and ECMWF data. Bull Am Meteorol Soc 80:1045–1075

Kaufman YJ, Koren I, Remer LA, Tanré D, Ginoux P, Fan S (2005) Dust transport and deposition observed from the terra-moderate resolution imaging spectroradiometer (MODIS) spacecraft over the Atlantic Ocean. J Geophys Res 110, doi:10,1029/2003JD004436

Kawahata H, Okamoto T, Matsumoto E, Ujiie H (2000) Fluctuations of eolian flux and ocean productivity in the mid-latitude North Pacific during the last 200 kyr. Quat Sci Rev 19:1279–1291

Kehl M, Frechen M, Skowronek A (2005) Paleosols derived from loess and loess-like sediments in the basin of Persepolis, southern Iran. Quat Int 140/141:135–149

Kellogg CA, Griffin DW, Garrison VH, Peak HK, Royall N, Smith RM, Shinn EA (2004) Characterization of aerosolized bacteria and fungi from desert dust events in Mali, West Africa. Aerobiologia 20:99–110

Kes AS (1983) Study of deflation processes and transfer of salts and dust. Probl Osvoen Pustin 1:3–15

Kes AS, Fedorovich BA (1976) Process of forming aeolian dust in space and time. Int Geogr Congr 23[Sect 1]:174–177

Khalaf FI, Gharib IM, Al-Kadi AS (1982) Source and genesis of the Pleistocene gravely deposits in northern Kuwait. Sediment Geol 3:101–117

Khiri F, Ezaidi A, Kabbachi K (2004) Dust deposits in Souss-Massa basin, south-west of Morocco: granulometrical, mineralogical and geochemical characterisation. J Afr Earth Sci 39:459–464

Kidson F, Gregory JW (1930) Australian origin of red rain in New Zealand. Nature 125:410–411

Kiefert L (1997) Characteristics of wind transported dust in Eastern Australia. PhD thesis, Faculty of Environmental Sciences, Griffith University, Griffith

Kiefert L, McTainsh GH (1996) Oxygen isotope abundance in the quartz fraction of Aeolian dust: implications for soil and ocean sediment formation in the Australasian region. Aust J Soil Res 34:467–473

Kim S-W, Yoon S-C, Jefferson A, Won J-G, Dutton EG, Ogren JA, Andreson TL (2004) Observation of enhamced water vapour in Asian dust layer and its effect on atmospheric radiative heating rates. Geophys Res Lett 13:L18113, doi:10.1029/2004GL020024

King MD, Kaufman YJ, Tanre D, Nakajima T (1999) Remote sensing of atmospheric aerosols from space: past, present and future. Bull Am Meteorol Soc 80:2229–2259

Kirkland TM, Fierer J (1996) Coccidioidomycosis: a reemerging infectious disease. Emerg Infect Dis 2:192–199

Kischa P, Barnaba F, Gobbi GP, Alpert P, Shtivelman A, Krichak SO, Joseph JM (2005) Vertical distribution of Saharan dust over Rome (Italy): comparison between 3-year model predictions and lidar soundings. J Geophys Res 110:D06028, doi:1029/2004JD005480

Kjelgaard J, Sharratt B, Sundram I, Lamb B, Claiborn C, Saxton K, Chandler D (2004) PM10 emission from agricultural soils on the Columbia Plateau: comparison of dynamic and time-integrated field-scale measurements and entrainment mechanisms. Agric For Meteorol 125:259–277

Klimenko LV, Moskaleva LA (1979) Frequency of occurrence of dust storms in the USSR. Meteorol Gidrol 9:93–97

Knight AW, McTainsh GH, Simpson RW (1995) Sediment loads in an Australian dust storm: implications for present and past dust processes. Catena 24:195–213

Knightle P (1993) Dust rain – 17 September 1992. Weather 48:58–59

Koçak M, Nimmo M, Kubilay N, Herut B (2004) Spatio-temporal aerosol trace metal concentrations and sources in the Levantine Basin of the Eastern Mediterranean. Atmos Environ 38:2133–2144

Koch J, Renno RO (2005) The role of convective plumes and vortices on the global aerosol budget. Geophys Res Lett 32:L18806

Kohfeld KE, Harrison SP (2003) Glacial–interglacial changes in dust deposition on the Chinese Loess Plateau. Quat Sci Rev 22:1859–1878

Kolla V, Biscaye PE (1977) Distribution and origin of quartz in the sediments of the Indian Ocean. J Sediment Petrol 47:642–649

Kolla VP, Biscaye PE, Hanley AF (1979) Distribution of quartz in Late Quaternary Atlantic sediments in relation to climate. Quat Res 11:261–277

Koren I, Kaufman YJ (2004) Direct wind measurements of Saharan dust events from Terra and Aqua satellites. Geophys Res Lett 31:L06122

Koren I, Joseph JH, Israelevich P (2003) Detection of dust plumes and their sources in northeastern Libya. Can J Remote Sensing 29:792–796

Kröhling DM (1999) Sedimentological maps of the typical loess units in North Pampa, Argentina. Quat Int 57/58:135–146

Kröhling DM (2003) A 54 m thick loess profile in North Pampa, Argentina. In: Geological Society of America Abstracts with Programs, p. 198

Krom MD, Cliff RA, Eijsink LM, Herut B, Chester R (1999) The characterisation of Saharan dusts and Nile particulate matter in surface sediments from the Levantine basin using Sr isotopes. Mar Geol 155:319–330

Kubilay N, Nickovic S, Moulin C, Dulac F (2000) An illustration of the transport and deposition of mineral dust onto the eastern Mediterranean. Atmos Environ 34:1293–1303

Kubilay N, Cokacar T, Oguz T (2003) Optical properties of mineral dust outbreaks over the northeastern Mediterranean. J Geophys Res 108:D21, doi: 10.1029/2003JD003798

Kubilay N, Oguz T, Koçak M, Torres O (2005) Ground-based assessment of total ozone mapping spectrometer (TOMS) data for dust transport over the northeastern Mediterranean. Global Biogeochem Cycles 19, doi:10.1029/2004GB002370

Kuenen PH (1960) Experimental abrasion 4: eolian action. J Geol 68:427–449

Kukal K (1971) Geology of recent sediments. Academic Press, London

Kukal K, Saadallah A (1973) Aeolian admixtures in the sediments of the northern Persian Gulf. In: Purser BH (ed) The Persian Gulf. Springer, Berlin Heidelberg New York, pp. 115–121

Kurosaki Y, Mikami M (2004) Effect of snow cover on threshold wind velocity of dust outbreak. Geophys Res Lett 31: L03106

Kurosaki Y, Mikami M (2005) Regional difference in the characteristic of dust event in East Asia: relationship among dust outbreak, surface wind, and land surface condition. J Meteorol Soc Jpn 83A:1–18

Kurtz AC, Derry LA, Chadwick OA (2001) Accretion of Asian dust to Hawaiian soils: isotopic, elemental and mineral mass balances. Geochim Cosmochim Acta 65:1971–1983

Kutiel H, Furman H (2003) Dust storms in the Middle East: sources of origin and their temporal characteristics. Indoor Built Environ 12:419–426

Kwaasi AAA, Parhar RS, Al-Mohanna FAA (1998) Aeroallergens and viable microbes in sandstorm dust – potential triggers of allergic and non-allergic respiratory ailments. Allergy 53:255–265

Kwon HJ, Cho SH, ChunY, Lagarde F, Pershagen G (2002) Effects of the Asian dust events on daily mortality in Seoul, Korea. Environ Res 90:1–5

Kyotani T, Koshimizu S, Kobayashi H (2005) Short-term cycle of aeolian dust (Kosa) recorded in Lake Kawaguchi sediments, central Japan. Atmos Environ 39:3335–3342

Laity JE (1994) Landforms of aeolian erosion. In: Abrahams AD, Parsons AJ (eds) Geomorphology of desert environments. Chapman and Hall, London, pp. 506–535

Lancaster N (1984) Characteristics and occurrence of wind erosion features in the Namib Desert. Earth Surf Process Landforms 9:469–478

Landis GA, Jenkins PP (2000) Measurement of the settling rate of atmospheric dust on Mars by the MAE instrument on Mars pathfinder. J Geophys Res 105:1855

Langston G, McKenna Neuman C (2005) An experimental study on the susceptibility of crusted surfaces to wind erosion: a comparison of the strength properties opf biotic and salt crusts. Geomorphology 72:40–53

Larrasoaña JC, Roberts AP, Rohling EJ, Winklhofer M, Wehausen R (2003) Three million years of monsoon variability over the northern Sahara. Clim Dyn 21:689–698

Larsen JA (1924) Dust storms of northern Idaho and western Montana. Mon Weather Rev 52:110

Lasserre F, et al. (2005) Development and validation of a simple mineral dust source inventory suitable for modelling in North central China. Atmos Environ 39:3831–3841

Lateef ASA (1988) Distribution, provenance, age and palaeoclimatic record of the loess in central North Iran. In: Eden DN, Furkert RJ (eds) Loess. Balkema, Rotterdam, pp. 93–101

Laurent B, Marticorena B, Bergametti G, Chazette P, Maigan F, Schmechtig C (2005) Simulation of the mineral dust emission frequencies from desert areas of China and Mongolia using an aerodynamic roughness length map derived from the POLDER/ADEOS 1 surface products. J Geophys Res 110:D, doi:10.1029/2004JD005013

Lavenu A, Fournier M, Sebrier, M (1984) Existence de nouveaux épisodes lacustres quaternaires dans l'Altiplano peruvo-bolivien. Cah ORSTOM Geol 14:103–114

Lazarenko AA (1984) The loess of Central Asia. In: Velichko A (ed) Late quaternary environments of the Soviet Union. Longman, Harlow, pp. 125–131

Leathers CR (1981) Plant components of desert dust in Arizona and their significance for man. In: Péwé TL (ed) Desert dust. Geol Soc Am Spec Pap 186:191–206

Le-Bolloch O, Guerzoni S, Molinaroli E (1996) Atmosphere–ocean mass fluxes at two coastal sites in Sardinia 39–41 degrees N, 8–10 degrees E. In: Guerzoni S, Chester R (eds) The impact of desert dust across the Mediterranean. Kluwer, Dordrecht, pp. 217–222

Lee JA, Wigner KA, Gregory JM (1993) Drought, wind and blowing dust on the southern High Plains of the United States. Phys Geogr 14:56–67

Lee JA, Allen BL, Peterson RE, Gregory JM, Moffett KE (1994) Environmental controls on blowing dust direction at Lubbock, Texas, USA. Earth Surf Process Landforms 19:437–449

Lee YI, Lim HS, Yoon HI (2004) Geochemistry of soils of King George Islands, West Antarctica: implications for pedogenesis in cold polar regions. Geochim Cosmochim Acta 68:4319–4333

Legrand M, Bertrand JJ, Desbois M, Meneger L, Fouquart Y (1989) The potential of satellite infrared satellite data for the retrieval of Saharan dust optical depth over Africa. J Appl Meteorol 28:309–318

Legrand M, Ndoume C, Jankowiak I (1994) Satellite derived climatology of the Saharan aerosol In: Lynch DK (ed) Passive infrared remote sensing of clouds and the atmosphere II. Proc SPIE 2309:127–35

Lehmkuhl F (1997) The spatial distribution of loess and loess-like sediments in the mountain areas of central and High Asia. Z Geomorphol Suppl 111:97–116

Lehmkuhl F, Klinge M, Rees-Jones J, Rhodes EJ (2000) Late Quaternary aeolian sedimentation in central and south-east Tibet. Quat Int 68/71:117–132

Leinen M, Heath GR (1981) Sedimentary indicators of atmospheric activity in the northern hemisphere during the Cenozoic. Palaeogeogr Palaeoclimatol Palaeoecol 36:1–21

Lenes JM, Darrow BP, Cattrall C (2001a) Iron fertilization and the *Trichodesmium* response on the West Florida shelf. Limnol Oceanogr 46:1261–1277

Lenes JM, et al. (2001b) Iron fertilization and the *Trichodesmium* response on the West Florida shelf. Limnol Oceanogr 46:1261–1277

Léon JF, LeGrand M (2003) Mineral dust sources in the surroundings of the North Indian Ocean. Geophys Res Lett 30:1309

Leovy C (2001) Weather and climate on Mars. Nature 412:245–249

Lepple FK, Brine CJ (1976) Organic constituents in eolian dust and surface sediments from north west Africa. J Geophys Res 81:1141–1147

Leroy M, et al. (1997) Retrieval of aerosol properties and surface bi-directional reflectances from POLDER/ADEOS. J Geophys Res 102:17023–17037

Leslie LM, Speer MS (2005) Changes in dust storm occurrence over central eastern Australia: 1950–2004. Proc. Int Symp Air Qual Manage Urban Reg Global Scales 3:3–12

Létolle R, Mainguet M (1993) Aral. Springer, Paris

Lever A, McCave IN (1983) Eolian components in Cretaceous and Tertiary North Atlantic sediments. J Sediment Petrol 53:811–832

Levin Z, Ganor E, Gladstein V (1996)The effects of desert particles with sulphate on rain formation in the Eastern Mediterranean. J Appl Meteorol 35:1511–1523

Li X, Maring H, Savoie D, Voss K, Prospero JM (1996) Dominance of mineral dust in aerosol light-scattering in the North Atlantic trade winds. Nature 380:416–419

Li X-Y, Liu L-Y (2003) Effect of gravel mulch on aeolian dust accumulation in the semiarid region of northwest China. Soil Tillage Res 70:73–81

Li X-Y, Liu L-Y, Gong J-D (2001) Influence of pebble mulch on soil erosion by wind and trapping capacity for windblown sediment. Soil Tillage Res 59:137–142

Li X-Y, Liu L-Y, Gao S-Y, Shi P-J, Zou X-Y, Hasi E, Yan P (2005) Aeolian dust accumulation by rock fragment substrata: influence of number and textural composition of pebble layers on dust accumulation. Soil Tillage Res 84:139–144

Liao H, Seinfeld JH, Adams PJ, Mickley LJ (2004) Global radiative forcing of coupled tropospheric ozone and aerosols in a unified general circulation model. J Geophys Res 109:D16207, doi:10.1029/2003JD004456

Liddell Hart BH (1953) The Rommel papers. Collins, London

Lim J, Matsumoto E, Kiatgawa H (2005) Eolian quartz flux variations in Cheju Island, Korea, during the last 6500 yr and a possible Sun–monsoon linkage. Quat Res 64:12–20

Lisitzin AP (1972) Sedimentation in the world ocean. Soc Econ Paleontol Mineral Spec Publ 17, 218 pp

Littmann T (1991a) Dust storm frequency in Asia: climatic control and variability. Int J Climatol 11:393–412

Littmann T (1991b) Rainfall, temperature and dust storm anomalies in the African Sahel. Geogr J 157:136–160

Littmann T (1991c) Recent African dust deposition in West Germany – sediment characteristics and climatological aspects. Catena Suppl 20:57–73

Liu T (1988) Loess in China. China Ocean Press, Beijing

Liu T, Ding Z (1998) Chinese loess and the paleomonsoon. Annu Rev Earth Planet Sci 26:111–145

Liu TS, Gu X, An ZS, Fan Y (1981) The dust fall in Beijing, China, on August 18th 1980. In: Péwé TL (ed) Desert dust. Geol Soc Am Spec Pap 186:149–157

Liu W, Feng Q, Wang T, Zhang Y, Shi J (2004) Physicochemistry and mineralogy and dust sediment in northern China. Adv Atmos Sci 21:775–783

Liu W, Yang H, Cao Y, Ning Y, Li L, Zhou J, An Z (2005) Did an extensive forest ever develop on the Chinese Loess Plateau during the last 130 ka?; a test using soil carbon isotopic signatures. Appl Geochem 20:519–527

Liu XM, Rolph T, An Z, Hesse P (2003) Paleoclimatic significance of magnetic properties on the Red Clay underlying the loess and paleosols in China. Palaeogeogr Palaeoclimatol Palaeoecol 199:153–166

Lockeretz W (1978)The lessons of the Dust Bowl. Am Sci 66:560–569

Loewe F (1943) Dust storms in Australia. Aust Meteorol Bur Bull 28

Logan J (1974) African dusts as a source of solutes in Gran Canaria ground waters. GSA Abstr Prog 6:849

López MV, Sabre M, Gracia R, Arrúe JL, Gomes L (1998) Tillage effects on soil surface conditions and dust emission by wind erosion in semiarid Aragón (NE Spain). Soil Tillage Res 45:91–105

Lourensz RS, Abe K (1983) A dust storm over Melbourne. Weather 38:272–275

Löye-Pilot MD, Martin JM (1996) Saharan dust input to the western Mediterranean: an eleven year record in Corsica. In: Guerzoni S, Chester R (eds) The impact of desert dust across the Mediterranean. Kluwer, Dordrecht, pp. 191–199

Löye-Pilot MD, Martin JM, Morelli J (1986) Influence of Saharan dust on the rainfall acidity and atmospheric input to the Mediterranean. Nature 321:427–428

Lu H, Sun D (2000) Pathways of dust input to the Chinese Loess Plateau during the last glacial and interglacial periods. Catena 40:251–261

Lu H, Vandenberge J, An Z (2001) Aeolian origin and palaeoclimatic implications of the "Red Clay" (north China) as evidenced by grain-size distribution. J Quat Sci 16:89–97

Lu H, Zhang F, Liu X, Duce RA (2004) Peridicities of palaeoclimatic variations recorded by loess-paleosol sequences in China. Quat Sci Rev 23:1891–1900

Luo C, Mahowald NM, Corral J del (2003) Sensitivity study of meteorological parameters on mineral aerosol mobilization, transport and distribution. J Geophys Res 108, doi:10.1029/2003JD003483

Lyell C (1834) Observations on the loamy deposit called "loess" of the basin of the Rhine. Edinburgh New Philos J 17:110–113, 118–120

Lyles L, Tatarko J, Dickerson JD (1983) Windbreak effects on soil water and wheat yield. Am Soc Agric Eng Pap 83:2074

Ma C-J, Kasahara M, Höller R, Kamiya T (2001) Characteristics of single particles sampled in Japan during the Asian dust-storm period. Atmos Environ 35:2707–2714

McCalla TM, Army TJ (1961) Stubble mulch farming. Adv Agron 13:125–196

McCauley JF, Grolier MJ, Breed CS (1977) Yardangs of Peru and other desert regions. US Geol Surv Interagency Rep Astrogeol 81, 177 pp

Mace KA, Kubilay N, Duce RA (2003) Organic nitrogen in rain and aerosol in the eastern Mediterranean atmosphere: an association with atmospheric dust. J Geophys Res 108: 4320

Macleod DA (1980) The origin of the red Mediterranean soils in Epirus, Greece. J Soil Sci 31:125–136

Magee JW, Miller GH (1998) Lake Eyre palaeohydrology from 60 ka to the present: Beach ridges and glacial maximum aridity. Palaeogeogr Palaeoclimatol Palaeoecol 144:307–329

Maher BA, Dennis PF (2001) Evidence against dust-mediated control of glacial–interglacial changes in atmospheric CO_2. Nature 411:176–180

Mahowald N, Dufresne JL (2004) Sensitivity of TOMS aerosol index to boundary layer height: implications for detection of mineral aerosol sources. Geophys Res Lett 31:L03103

Mahowald N, Luo C (2003) A less dusty future? Geophys Res Lett 30, doi: 10.1029/2003GL017880

Mahowald N, Kohfeld K, Hansson M, Balkanski Y, Harrison SP, Prentice IC, Schulz, M, Rodhe H (1999) Dust sources and deposition during the last glacial maximum and current climate: a comparison of model results with paleodata from ice cores and marine sediments. J Geophys Res 104:15895–15916

Mainguet M (1972) Le Modélé des Grès. IGN, Paris

Mainguet M (1980) Aeolian interdependencies of arid Saharan areas on the borders of the Sahel and the consequences for the propagation of desertification. Stuttgart Geogr Stud 95:107–123

Mainguet M, Chemin M-C (1990) Le Massif du Tibesti dans le système éolien du Sahara: reflexion sur la génèse du Lac Tchad. Berlin Geogr Stud 30:261–276

Maley J (1982) Dust, clouds, rain types and climatic variations in tropical north Atlantic. Quat Res 18:1–16

Malik JN, Khadikikar AS, Merh SS (1999) Allogenic control on late Quaternary continental sedimentation in the Mahi river basin, Western India. J Geol Soc India 53:299–314

Malin JC (1946) Dust storms. Kansas Hist Q 14:129–133, 265–296, 391–413

Marchand DE (1970) Soil contamination in the White Mountains, eastern California. Bull Geol Soc Am 81:2497–2506

Maring XLH, Savoie KV, Prospero JM (1996) Dominance of mineral dust in aerosol light-scattering in the North Atlantic trade winds. Nature 380:416–419

Marticorena B, Bergametti G (1996) Two-years simulations of seasonal and inter-annual changes of the Saharan dust emissions. Geophys Res Lett 23:1921–1924

Marticorena B, Bergametti G, Legrand M (1999) Comparison of emission models used for large scale simulation of the mineral dust cycle. Contrib Atmos Phys 72:151–160

Martin JH, Gordon RM, Fitzwater SE (1991) The case for iron. Limnol Oceanogr 36:1793–1802

Martin RJ (1937) Dust storms of January–April 1937 in the United States. Mon Weather Rev 65:151–152

Marx S, McGowan HA (2005) Dust transportation and deposition in a superhumid environment, West Coast, South Island, New Zealand. Catena 59:147–171

Marx SK, Kamber BS, McGowan HA (2005a) Estimates of Australian dust flux into New Zealand: quantifying the eastern Australian dust plume pathway using trace element calibrated ^{210}Pb as a monitor. Earth Planet Sci Lett 239:336–351

Marx S, Kamber BS, McGowan HA (2005b) Provenance of long-travelled dust determined with ultra-trace-element composition: a pilot study with samples from New Zealand glaciers. Earth Surf Process Landforms 30:699–716

Masmoudi M, Chaabane M, Medhioub K, Elleuch F (2003) Variability of aerosol optical thickness and atmospheric turbidity in Tunisia. Atmos Res 66:175–188

Masdon JA, Kuzila MS (2000) Episodic Holocene loess deposition in central Nebraska. Quat Int 67:119–131

Mason JA (2001) Transport direction of Peoria Loess in Nebraska and implications for loess sources on the Central Great Plains. Quat Res 56:79–86

Mason JA, Jacobs PM, Hanson PR, Miao X, Goble RJ (2003) Sources and paleoclimatic significance of Holocene Bignell loess, central Great Plains, USA. Quat Res 60: 330–339

Masri Z, Zöbisch M, Bruggeman A, Hayek P, Kardous M (2003) Wind erosion in a marginal Mediterranean dryland area: a case study from the Khanasser Valley, Syria. Earth Surf Process Landforms 28:1211–1222

Matson M, Holben B (1987) Satellite detection of tropical burning in Brazil. Int J Remote Sensing 8:509–516

Mattsson JO, Nihlén T (1996) The transport of Saharan dust to southern Europe: a scenario. J Arid Environ 32:111–110

Mayer L, McFadden LD, Harden JW (1988) Distribution of calcium carbonate in desert soils: a model. Geology 16:303–306

McCalla TM, Army TJ (1961) Stubble mulch farming. Adv Agron 13:125–196

McCauley JF, Grolier MJ, Breed CS (1977) Yardangs of Peru and other desert regions. US Geol Surv Interagency Rep Astrogeol 81, 177 pp

McCauley JF, Breed CS, Grolier MJ, MacKinnon DJ (1981) The US dust storms of February 1977. In: Péwé TL (ed) Desert dust. Geol Soc Am Spec Pap 186:123–148

McDonald WF (1938) Atlas of climatic charts of the oceans. Department of Agriculture, Weather Bureau, Washington, D.C.

McFadden LD, Wells SG, Jercinovich MJ (1987) Influence of eolian and pedogenic processes on the origin and evolution of desert pavements. Geology 15:504–508

McGowan HA, McTainsh GH, Zawar-Reza P, Sturman AP (2000) Identifying regional dust transport pathways: application of kinematic trajectory modelling to a trans-Tasman case. Earth Surf Process Landforms 25:633–647

McGowan HA, Kamber B, McTainsh GH, Marx SK (2005) High resolution provenancing of long travelled dust deposited on the Southern Alps, New Zealand. Geomorphology 69:208–221

McKendry IG, Hacker JP, Stull R, Sakiyama S, Mignacca D, Reid K (2001) Long-range transport of Asian dust to the Lower Fraser Valley, British Columbia, Canada. J Geophys Res 106:18361–18370

McKenna M, Neuman C, Maxwell CD, Bouton JW (1996) Wind transport of sand surface crusted with photoautrophic micro organisms. Catena 27:229–247

McMahon H (1906) Recent survey and exploration in Seistan. Geogr J 28:209–228

McTainsh GH (1980) Harmattan dust deposition in northern Nigeria. Nature 286:587–588

McTainsh GH (1984) The nature and origin of Aeolian mantles in northern Nigeria. Geoderma 33:13–37

McTainsh GH (1985) Dust processes in Australia and West Africa: a comparison. Search 16:104–106

McTainsh GH (1987) Desert loess in northern Nigeria. Z Geomorphol NF 26:417–435

McTainsh GH (1989) Quaternary aeolian dust processes and sediments in the Australian region. Quat Sci Rev 8:235–253

McTainsh GH (1999) Dust transport and deposition. In: Goudie AS, Livingstone I, Stokes S (eds) Aeolian environments, sediments and landforms. Wiley, Chichester, pp. 181–211

McTainsh GH, Pitblado JR (1987) Dust storms and related phenomena measured from meteorological records in Australia. Earth Surf Process Landforms 12:415–424

McTainsh GH, Walker PH (1982) Nature and distribution of Harmattan dust. Z Geomorphol 26:417–435

McTainsh GH, Burgess R, Pitblado JR (1989) Aridity, drought and dust storms in Australia (1960–84). J Arid Environ 16:11–22

McTainsh GH, Lynch AW, Burgess RC (1990) Wind erosion in eastern Australia. Aust J Soil Res 28:323–339

McTainsh GH, Leys J, Nickling WG (1999) Wind erodibility of arid lands in the Channel Counry of western Queensland, Australia. Z Geomorphol Suppl 116:113–130

McTainsh GH, Chan Y-C, McGowan H, Leys J, Tews K (2005) The 23rd October 2002 dust storm in eastern Australia: characteristics and meteorological conditions. Atmos Environ 39:1227–1236

Measures CI, Brown ET (1996) Estimating dust input to the Atlantic Ocean using surface water aluminium concentrations. In: Guerzoni S, Chester R (eds) The impact of desert dust across the Mediterranean. Kluwer, Dordrecht, pp. 301–311

Melis MI, Acworth RI (2001) An Aeolian component in Pleistocene and Holocene valley aggradation: evidence from Dicks Creek catchment, Yass, New South Wales. Aust J Soil Res 39:13–38

Meloni D, Di Sarra A, Di Iorio T, Fiocco G (2005) Influence of the vertical profile of Saharan dust on the visible direct radiative forcing. J Quant Spectrosc Radiat Transfer 93:397–413

Membery DA (1983) Low level wind profiles during the Gulf Shamal. Weather 38:18–24

Merrill SD, Black AL, Fryrear DW, Saleh A, Zobeck TM, Halvorson AD, Tanaka DL (1999) Soil wind erosion hazard of spring wheat-fallow as affected by long-term climate and tillage. J Soil Sci Soc Am 63:1768–1777

Meskhidze N, Chameides WL, Nenes A (2005) Dust and pollution: a recipe for enhanced ocean fertilization? J Geophys Res 110, doi:10.1029/2004JD005082

Mestdagh H, Haesaert P, Dodonov A, Hus J (1999) Pedosedimentary and climatic reconstruction of the last interglacial and early glacial loess-paleosol sequence in South Tadzhikistan. Catena 35:197–218

Mezösi G, Szatmári J (1998) Assessment of wind erosion risk on the agricultural area of the southern part of Hungary. J Hazard Mat 61:139–153

Miao X, Mason JA, Goble RJ, Hanson PR (2005) Loess record of dry climate and Aeolian activity in the early- to mid-Holocene central Great Plains, North America. Holocene 15:339–346

Michaelides S, Evripidou P, Kallos G (1999) Monitoring and predicting Saharan Desert dust events in the eastern Mediterranean. Weather 54:359–365

Michels K, Sivakumar MVK, Allison BE. (1993) Wind erosion in the southern Sahelian zone and induced constraints to pearl millet production. Agric For Meteorol 67:65–77

Michels K, Sivakumar MVK, Allison BE (1995) Wind erosion control using crop residue: I. Effects on soil flux and soil properties. Field Crops Res 40:101–110

Micklin PP (1988) Desiccation of the Aral Sea: a water management disaster in the Soviet Union. Science 241:1170–1176

Middleton NJ (1984) Dust storms in Australia: frequency, distribution and seasonality. Search 15:46–47

Middleton NJ (1985a) Effect of drought on dust production in the Sahel. Nature 316:431–434

Middleton NJ (1985b) Dust production in the Sahel: reply to M. Hulme. Nature 318:488

Middleton NJ (1986a) Dust storms in the Middle East. J Arid Environ 10:83–96

Middleton NJ (1986b) A geography of dust storms in south-west Asia. J Climatol 6:183–196

Middleton NJ (1986c) The geography of dust storms. DPhil thesis, University of Oxford, Oxford

Middleton NJ (1989) Climatic controls on the frequency, magnitude and distribution of dust storms: examples from India and Pakistan, Mauritania and Mongolia. In: Leinen M, Sarnthein M (eds) Palaeoclimatology and palaeometeorology: modern and past patterns of global atmospheric transport. NATO ASI Ser C 282:97–132

Middleton NJ (1990) Wind erosion and dust storm prevention. In: Goudie AS (ed) Desert Reclamation. Wiley, Chichester, pp. 87–108

Middleton NJ (1991) Dust storms in the Mongolian People's Republic. J Arid Environ 20:287–297

Middleton NJ (2002) The Aral Sea. In: Shahgedanova M (ed) The physical geography of Northern Eurasia. Oxford University Press, Oxford, pp. 497–510

Middleton NJ, Chaudhary QZ (1988) Severe dust storm at Karachi, 31 May 1986. Weather 43:298–301

Middleton NJ, Goudie AS (2001) Saharan dust: sources and trajectories Trans Inst Br Geogr NS 26:165–181

Middleton NJ, Thomas DSG (1997) World atlas of desertification. Arnold, London

Middleton NJ, Goudie AS, Wells GL (1986) The frequency and source areas of dust storms. In: Nickling WG (ed) Aeolian geomorphology. Allen and Unwin, Boston, pp. 237–259

Middleton NJ, Betzer PR, Bull PA (2001) Long-range transport of 'giant' aeolian quartz grains: linkage with discrete sedimentary sources and implications for protective particle transfer. Mar Geol 177:411–417

Mignon C, Sandroni V (1999) Phosphorus in rainwater: partitioning inputs and impact on the surface coastal ocean. Limnol Oceanogr 44:1160–1165

Mill HR (1902) Dust showers in the south-west of England. Symons Mon Meteorol Mag 37:1–4

Mill HR, Lempfert RGK (1904) The great dust-fall of February 1903, and its origin. Q J R Meteorol Soc 30:57–91

Miller RL, Tegen I (1998) Climate response to soil dust aerosols. J Clim 11:3247–3267

Miller RL, Tegen I, Perlwitz J (2004a) Surface radiative forcing by soil dust aerosols and the hydrological cycle. J Geophys Res 109D, doi: 10.1029/2003JD004085

Miller RL, Perlwitz J, Tegen I (2004b) Modelling Arabian dust mobilization during the Asian summer monsoon: the effect of prescribed versus calculated SST. Geophys Res Lett 31:L22214, doi:10.1029/2004GL20669

Mills MM, Ridame C, Davey M, La Roche J, Gelder RJ (2004) Iron and phosphorus co-limit nitrogen fixation in the eastern tropical North Atlantic. Nature 429:292–294

Mitchell JM, Stockton CW, Meko DM (1979) Evidence of a 22-year rhythm of drought in the western United States related to Hale solar cycle since the 17th century. In: McCormac BM, Seliga TA (eds) Solar–terrestrial influences on weather and climate. Reidel, Dordrecht, pp. 125–144

Mo KC, Higgins RW (1996) Large-scale atmospheric moisture transport as evaluated in the NCEP/NCAR and NASA/DAO reanalyses. J Clim 9:1531–1545

Modaihsh AS (1997) Characteristics and composition of the falling dust sediments on Riyadh City, Saudi Arabia. J Arid Environ 36:211–223

Mohammed AB, Frangi J-P (1986) Results from ground-based monitoring of spectral aerosol optical thickness and horizontal extinction: some specific characteristics of dusty Sahelian atmospheres. J Clim Appl Meteorol 25:1807–1815

Mohsin SI, Farooqui MA, Danish M (1989) A textural study of nearshore recent sediments, Makran coast, Pakistan. Pak J Sci Ind Res 32:13–16

Molesworth AM, Cuevas LE, Morse AP (2002) Dust clouds and spread of infection. Lancet 359:81–82

Molinaroli E (1996) Mineralogical characterisation of Saharan dust with a view to its final destination in Mediterranean sediments. In: Guerzoni S, Chester R (eds) The impact of desert dust across the Mediterranean. Kluwer, Dordrecht, pp. 153–162

Møller JT (1986) Soil degradation in a north European region. In: Fantechi R Margaris NS (eds) Desertification in Europe. Reidel, Dordrecht, pp. 214–230

Molosnova TI, Subbotina OI, Chanysheva SG (1987) Climatic consequences of economic activity in the zone of the Aral Sea (in Russian). Gidrometeoizdat, Moscow, 119 pp

Monteil MA (2002) Dust clouds and spread of infection. Lancet 359:81

Moon AE (1982) Dust-fall at Hastings, Sussex. J Meteorol 7:92

Morales AF (1946) El Sahara Español. Alta Comisaria de España en Marruecos, Madrid

Morales C (ed) (1979) Saharan dust, mobilization, transport, deposition. Wiley, Chichester

Moreno A, Canals M (2004) The role of dust in abrupt climate change: insights from offshore northwest Africa and Alboran Sea sediment records. Contrib Sci (Barcelona) 2:485–498

Moreno A, Targarona J, Henderiks J, Canals M, Freudenthal T, Meggers H (2001) Orbital forcing of dust supply to the North Canary Basin over the last 250 kyr. Quat Sci Rev 20:1327–1339

Moreno A, Cacho I, Canals M, Prins MA, Sanchéz-Goñi M-F, Grimalt JO, Weltje GJ (2002) Saharan dust transport and high-latitude glacial climatic variability: the Alboran Sea record. Quat Res 58:318–328

Morgan M (2000) Sting of the scorpion. Sutton, Stroud

Morgan RPC (1995) Soil erosion and conservation, 2nd edn. Longman, Harlow

Mori I, Nishikawa M, Quan H, Morita M (2002) Estimation of the concentration and chemical composition of kosa aerosols at their origin. Atmos Environ 36:4569–4575

Mori I, Nishikawa M, Tanimura T, Hao Q (2003) Change in size distribution and chemical composition of kosa (Asian dust) aerosol during long-range transport. Atmos Environ 37:4253–4263

Mottershead DN, Pye K (1994) Tafoni on coastal slopes, South Devon, UK. Earth Surf Process Landforms 19:543–563

Moulin C, Chiapello I (2003) Evidence of the control of summer atmospheric transport of African dust over the Atlantic by Sahel sources from TOMS satellites (1979–2000). Notes Act Instrum IPSL 40, 12 pp

Moulin C, Chiapello I (2004) Evidence of the control of summer atmospheric transport of African dust over the Atlantic by Sahel sources from TOMS satellites. Geophys Res Lett 31:L02107

Moulin C, Guillard F, Dulac F, Lambert CE (1997a) Long-term daily monitoring of Saharan dust load over ocean using Meteosat ISCCP-B2 data, 1: methodology and preliminary results for 1983–1994 in the Mediterranean. J Geophys Res 102:16947–16958

Moulin C, Lambert C E, Dulac F, Dayan U (1997b) Control of atmospheric export of dust by the North Atlantic Oscillation. Nature 387:691–694

Muhs DR, Bettis EA (2000) Geochemical variations in Peoria loess of western Iowa indicate paleowinds of midcontinental North America during last glaciation. Quat Res 53:49–61

Muhs DR, Maat PB (1993) The potential response of eolian sands to greenhouse warming and precipitation reduction on the Great Plains of the United States. J Arid Environ 25:351–361

Muhs DR, Bush CA, Stewart KC (1990a) Geochemical evidence of Saharan dust parent material for soils developed on Quaternary limestones of Caribbean and western Atlantic islands. Quat Res 33:157–177

Muhs DR, Bush CA, Stewart KC, Rowland TR, Crittenden RC (1990b) Geochemical evidence of Saharan dust parent material for soils developed on Quaternary limestones on Caribbean and Western Atlantic Islands. Quat Res 33:157–177

Muhs DR, Swinehart JB, Loope DB, Aleinikoff JN, Been J (1999) 200,000 years of climate change recorded in eolian sediments of the High Plains of eastern Colorado and western Nebraska. Geol Soc Am Field Guide 1:71–91

Muhs DR, McGeehin JP, Beann J, Fisher E (2004) Holocene loess deposition and soil formation as competing processes, Matanuska Valley, southern Alaska. Quat Res 61:265–276

Mullen RE, Darby DA, Clark DL (1972) Significance of atmospheric dust and ice rafting for Arctic Ocean sediment. Bull Geol Soc Am 83:205–212

MWR (1980) Dustfall and purple sky in Berlin, 29 November 1979. J Meteorol 5:156–157

Nahon D, Trompette R (1982) Origin of siltstones: glacial grinding versus weathering. Sedimentology 29:29–35

Naiman Z, Quade J, Patchett PJ (2000) Isotopic evidence for eolian recycling of pedogenic carbonate and variations in carbonate dust sources throughout the southwest United States. Geochem Cosmochim Acta 64:3099–3109

Nakano T, Yokoo Y, Nishikawa M, Koyanagi H (2004) Regional Sr-Nd isotopic ratios of soil minerals in northern China as Asian dust fingerprints. Atmos Environ 38:3061–3067

Nakano T, Nishikawa M, Mori I, Shin K, Hosono T, Yokoo Y (2005) Source and evolution of the "perfect Asian dust storm" in ealy April 2001: implications of the Sr-Nd isotope ratios. Atmos Environ 39:5568–5575

Nalivkin DV (1983) Hurricanes, storms and tornadoes. Amerind, New Delhi

Naruse T, Sakai H, Inoue K (1986) Aeolian dust origin of fine dust in selected soils. Jpn Quat Res 24:295–300

Nash DJ, Thomas DSG, Shaw PA (1994) Timescales, environmental change and dryland valley development. In: Millington AC, Pye K (eds) Environmental change in drylands. Wiley, Chichester, pp. 25–41

Natsagdorj DL, Jugder D, Chung YS (2003) Analysis of dust storms observed in Mongolia during 1937–1999. Atmos Environ 37:1401–1411

Naval Intelligence Division (1943) Algeria (Geographical handbook series, vol 1). OUP, Oxford, 313 pp

Ndoye M, Gahukar R (1987) Insect pests of pearl millet in West Africa and their control. In: ICRISAT (ed) Proceedings International Pearl Millet Workshop, ICRISAT Center, Patancheru, pp. 195–205

Negi BS, Sadasivan S, Nambi KSV, Pande BM (1996) Characterization of atmospheric dust at Gurushikar, Mt Abu, Rajasthan. Environ Monit Assessment 40/43:253–259

Nettleton WD, Chadwick OA (1996) Late Quaternary redeposited loess-soil developmental sequences, South Yemen. Geoderma 70:21–36

Neuer S, Torres-Padron ME, Galado-Caballero MD, Rueda MJ, Hernández-Brito J, Davenport R, Wefer G (2004) Response of upper coean biogeochemistry to dust deposition pulses in the eastern subtropical North Atlantic gyre at time series station ESTOC. Global Biogeochem Cycles 18:GB 4020, doi: 10.1029/2004GB002228

Newell RE, Kidson JW (1984) African mean wind changes between Sahelian wet and dry periods. J Climatol 4:27–33

Nicholson SE (1986) The spatial coherence of African rainfall anomalies: inter-hemispheric teleconnections. J Clim Appl Meteorol 25:1365–1381

Nicholson SE (2000) Land surface processes and Sahel climate. Rev Geophys 38:117–139

Nickling WG (1978) Eolian sediment transport during dust storms: Slims River Valley, Yukan Territory. Can J Earth Sci 15:1069–1084

Nickling WG, Brazel AJ (1984) Transport and spatial characteristics of Arizona dust storms (1965–1980). J Climatol 4:645–660

Nickling WG, Gillies JA (1989) Emission of fine grained particulates from desert soils. In: Leinen M, Sarnthein M (eds) Paleoclimatology and palaeometeorology: modern and past patterns of global atmospheric transport. Kluwer, Dordrecht, pp. 133–166

Nickling WG, Gillies JA (1993) Dust emission and transport in Mali, West Africa. Sedimentology 40:859–868

Nickling WG, McTainsh GH, Leys JF (1999) Dust emissions from the Channel Country of western Queensland, Australia. Z Geomorphol Suppl 116:1–17

Nihlen T, Olsson S (1995) Influence of eolian dust on soil formation in the Aegean area. Z Geomorphol 39:341–261

Nilson E, Lehmkuhl F (2001) Interpreting temporal patterns in the late Quaternary dust flux from Asia to the North Pacific. Quat Int 76/77:67–76

Nordstrom KF, Hotta S (2004) Wind erosion from cropland in the USA: a review of problems, solutions and prospects. Geoderma 121:157–167

N'Tchayi GM, Bertrand J, Legrand M, Baudet J (1994). Temporal and spatial variations of the atmospheric dust loading throughout West Africa over the last thirty years. Ann Geophys 12:265–273

N'Tchayi GM, Bertrand JJ, Nicholson SE (1997) The diurnal and seasonal cycles of wind-borne dust over Africa north of the Equator. J Appl Meteorol 36:868–882

Nutgeren G, Vandenberghe J, Huissteden JK van, An Z (2004) A Quaternary climate record based on grain size analysis from the Luochuan loess section on the central Loess Plateau, China. Global Planet Change 41:167–183

O'Carroll B (ed) (2003) Bearded brigands. Leo Cooper, Barnsley.

Offer ZY, Azmon E (1994) Chemistry and mineralogy of four dust storms in the northern Negev Desert, Israel (1988–1992). Science of the Total Environment 143:235–243

Offer ZY, Goossens D (2001) Ten years of Aeolian dust dynamics in a desert region (Negev desert, Israel): analysis or airborne dust concentration, dust accumulation and the high-magnitude dust events. J Arid Environ 47:211–249

O'Hara S, Wiggs G, Mamedov B, Davidson G, Hubbard RB (2000) Exposure to airborne dust contaminated with pesticide in the Aral Sea region. Lancet 355:627–628

Okin GS, Mahowald N, Chadwick OA, Artaxo P (2004) Impact of desert dust on the biogeochemistry of phosphorus in terrestrial ecosystems. Global Biogeochem Cycles 18:GB 2005

Okhravi R, Amini A (2001) Characteristics and provenance of the loess deposits of the Gharatikan watershed in Northeast Iran. Global Planetary Change 28:11–22

Olbruck G (1973) Staubfälle. Seewart 34:242–249

Oldfield F, Hunt A, Jones MDH, Chester R, Dearing JA, Olsson L, Prospero JM (1985) Magnetic differentiation of atmospheric dusts. Nature 317:516–518

Oliver FW (1945) Dust storms in Egypt and their relation to the war period, as noted in Maryut 1939–45. Geogr J 106:26-49

Orgill MM, Sehmel GA (1976) Frequency and diurnal variation of dust storms in the contiguous U.S.A. Atmos Environ 10:813–825

Orlovsky N, Orlovsky L (2001) White sandstorms in Central Asia. In: Yang Y, Squires, V, Lu Q (eds) Global alarm: dust and sandstorms from the world's drylands. United Nations, Bangkok, pp. 169–201

Orlovsky L, Orlovsky N, Durdyev A (2005) Dust storms in Turkmenistan. J Arid Environ 60:83–97

Osada K, Ilda H, Kido M, Matsunaga K, Iwasaka Y (2004) Mineral dust layers in snow at Mount Tateyama, central Japan: formation processes and characteristics. Tellus 56B:382–392

Overpeck J, Rind D, Lacis A, Healy R (1996) Possible role of dust-induced regional warming in abrupt climate change during the last glacial period. Nature 384:447–449

Ozer P, Erpicum M, Cortemiglia GC, Luchettti G (1998) A dustfall event in November 1996 in Genoa, Italy. Weather 53:140–145

Özsoy T, Saydam AC (2000) Acidic and alkaline precipitation in the Cilician Basin, north-eastern Mediterranen Sea. Sci Total Environ 253:93–109

Page LR, Chapman RW (1934) The dustfall of December 15;-16, 1933. Am J Sci (Ser 5) 28:288–297

Palmer AS, Pillans BJ (1996) Record of climatic fluctuations from ca.500 ka loess deposits and paleosols near wanganui, New Zealand. Quat Int 34/36:155–162

Pankine AA, Ingersoll AP (2004) Interannual variability of Mars global dust storms: an example of self-organized criticality? Icarus 170:514–518

Pant RK, Basavaiah N, Juyal N, Saini NK, Yadava MG, Appel E, Singhvi AK (2005) A 20-ka climate record from Central Himalayn loess deposits. J Quat Sci 20:485–492

Papayannis A, et al. (2005) First systematic observations of Saharan dust over Europe (2000–2003): statistical analysis and results. Geophys Res Abstr 7:04016

Paquet H, Coudé-Gaussen G, Rognon P (1984) Etude minéralogique de poussières sahariennes le long d'un itinéraire entre 19° et 35° de latitude nord. Rev Geol Dyn Geogr Phys 25:257–265

Pauley PM, Baker NL, Barker EH (1996) An observational study of the "Interstate 5" dust storm case. Bull Am Meteorol Soc 77:693–720

Park S-U, Chang L-S, Lee E-H (2005) Direct radiative forcing due to aerosols in East Asia during a Hwangsa (Asian Dust) event observed on 19–23 March 2002 in Korea. Atmos Environ 39:2593–2606

Parkin DW, Phillips DR, Sullivan DAR (1970) Airborne dust collections over the North Atlantic. J Geophys Res 75:1782–1793

Parungo F, Li Z, Li X, Yang D, Harris J (1994) Gobi dust storms and The Great Green Wall. Geophys Res Lett 21:999–1002

Patterson DB, Farley KA (1997) 4He as a tracer of Aeolian dust: a two million year record from ODP Site 806, Ontong Java Plateau. Annu VM Goldschmidt Conf 7:pdf 2200

Pease PP, Tchakerian VP, Tindale NW (1998) Aerosols over the Arabian Sea: geochemistry and source areas for aeolian desert dust. J Arid Environ 29:477–496

Peck JA, Green RR, Shanahan T, King JW, Overpeck JT, Scholz CA (2004) A magnetic mineral record of Late Quaternary tropical climate variability from Lake Bosumtwi, Ghana. Palaeogeogr Palaeoclimatol Palaeoecol 215:37–57

Pedersen TF, Bertrand P (2000) Influences of oceanic rheostats and amplifiers on atmospheric CO2. Quat Sci Rev 19:273–283

Pelletier JD, Cook JP (2005) deposition of playa windblown dust over geologic time scales. Geology 33:909–912

Perkins D (2004) Saharan dustfalls on Anglesey in December 2003: first confirmed depositions in December in the British Isles. Available at: http://www.llansadwrn-wx.co.uk/watch/dustdec03.html (accessed March 2005)

Perry KD, Cahill TA, Eldred RA, Dutcher DD (1997) Long-range transport of North African dust to the eastern United States. J Geophys Res 102:11225–11238

Pesci M (1968) Loess. In: Fairbridge RW (ed) The encyclopaedia of geomorphology. Reinhold, New York, pp. 674–678

Peterson JT, Junge CE (1971) Source of particulate matter in the atmosphere. In: Matthews W, Kellog W, Robinson GD (eds) Man's impact on climate. MIT Press, Boston, pp. 310–320

Petit JR, et al. (1999) Climate and atmospheric history of the past 420,000 years from the Vostok ice core, Antarctica. Nature 399:429–436

Petit RH, Legrand M, Jackowiak I, Molinié J, de Beauville CA, Marion G, Mansot JL (2005) Transport of Saharan dust over the Caribbean Islands: study of an event. J Geophys Res 110(D), doi:10.1029/2004JD004748

Petit-Maire N (ed) (1991) Paléoenvironments du Sahara. Editions du Centre National de la Recherche Scientific, Paris

Pettke T, Halliday AN, Hall CM, Rea DK (2000) Dust production and deposition in Asia and the North Pacific Ocean over the past 12 Myr. Earth Planet Sci Lett 178:397–413

Péwé TL (1951) An observation on wind-blown silt. J Geol 59:399–401

Péwé TL, Péwé EA, Péwé RH, Journaux A, Slatt RM (1981) Desert dust: characteristics, and rates of deposition in Central Arizona. Geol Soc Am Spec Pap 186: 169–190

Phillips RE, Blevins RL, Thomas GW, Frye WW, Phillips SH (1980) No-tillage agriculture. Science 208:1108–1113

Pias J (1971) Les loess en Afghanistan Oriental et leurs pédogenèses successives au Quaternaire recent. C R Acad Sci Paris 272(D):1602–1605

Pichevin L, Cremer M, Giraudeau J, Bertand P (2005) A 190 ky record of lithogenic grain-size on the Namibian slope: forging a tight link between past wind-strength and coastal upwelling dynamics. Mar Geol 218:81–96

Pierangelo C, Chédin A, Heilliette S, Jacquinet-Husson N, Armante R (2004) Dust altitude and infrared optical depth from AIRS. Atmos Chem Phys 4:1813–1822

Pike WS (1984) Dust-fall, west Berkshire, 24 September 1983. J Meteorol 9:21–22

Pinker RT, Pandithurai G, Holben BN, Dubovik O, Aro TO (2001) A dust outbreak episode in sub-Sahel West Africa. J Geophys Res 106(D19):22923–22930

Pisano G, Spinelli N, D'Avano L, Boselli A, Wang X, Papayannis A (2005) Atmospheric African mineral dust monitoring with Raman Lidar over Napoli. Geophys Res Abstr 7:09752

Pitty AF (1968) Particle size of the Saharan dust which fell in Britain in July 1968. Nature 220:364–365

Plumb RC, Tantayanon R, Libby M, Wen WX (1989) Chemical model for Viking biology experiments: implications for the composition of the Martian regolith. Nature 338:633–635

Pokras EM (1989) Pliocene history of South Saharan/Sahelian aridity: record of freshwater diatoms (genus *Melosira*) and opal phytoliths, ODP sites 662 and 664. In: Leinen M, Sarnthein M (eds) Palaeoclimatology and palaeometeorology: modern and past patterns of global atmospheric transport. Kluwer, Dordrecht, pp. 795–804

Porter SC (2001) Chinese loess record of monsoon climate during the last glacial–interglacial cycle. Earth Sci Rev 54:115–128

Porter SC, An Z (2005) Episodic gullying and paleomonsoon cycles on the Chinese Loess Plateau. Quat Res 64:234–241

Pourmand A, Marcantonio F, Schulz H (2004) Variations in productivity and eolian fluxes in the northeastern Arabian Sea during the past 110 ka. Earth Planet Sci Lett 221:39–54

Prigent C, Tegen I, Aires F, Marticorena B, Zribi M (2005) Estimation of the aerodynamic roughness length in arid and semi-arid regions over the globe with the ERS scatterometer. J Geophys Res 110:D09205, doi:10.1029/2004JD005370

Pringle AW, Bain DC (1981) Saharan dustfalls on north-west England. Geogr Mag 53:729–732

Prins MA, Postma G, Weltje GJ (2000) Controls on terrigenous sediment supply to the Arabian Sea during the later Quaternary: the Makran continental slope. Mar Geol 169:351–371

Prinz EM, Wenzel WP (1992) Geostationary satellite detection of biomass burning in South America. Int J Remote Sensing 13:2783–2799

Prodi F, Fea G (1979) A case of transport and deposition of Saharan dust over the Italian peninsula and Southern Europe. J Geogr Res 84:6951–6960

Prospero JM (1981a) Eolian transport to the world ocean. In: Emiliani C (ed) The sea: the oceanic lithosphere. John Wiley and Sons, New York, pp. 801–874

Prospero JM (1981b) Arid regions as sources of mineral aerosols in the marine atmosphere. In: Péwé TL (ed.) Desert dust. Geol Soc Am Spec Pap 186:71–86

Prospero JM (1996a) Saharan dust transport over the North Atlantic Ocean and Mediterranean: an overview. In: Guerzoni S, Chester R (eds) The impact of desert dust across the Mediterranean. Kluwer, Dordrecht, pp. 133–151

Prospero JM (1996b) The atmospheric transport of particles to the ocean. In: Ittekkot V, Schafer P, Honjo S, Depetris PJ (eds) Particle flux in the ocean. Wiley, Chichester, pp. 19–52

Prospero JM (1999) Long-term measurements of the transport of African mineral dust to the south-eastern United States: implications for regional air quality. J Geophys Res 104(D13):15917–15927

Prospero JM (2004) Interhemispheric transport of viable fungi and bacteria from Africa to the Caribbean with soil dust. In: Werner D (ed) Biological resources and their migration. Springer, Berlin Heidelberg New York, pp. 127–132

Prospero JM, Bonatti E (1969) Continental dust in the atmosphere of the eastern equatorial Pacific. J Geophys Res 74:3362–3371

Prospero JM, Carlson TN (1972) Vertical and areal distribution of Saharan Dust over the western equatorial North Pacific Ocean. J Geophys Res 77:5255–5265

Prospero JM, Carlson TN (1981) Saharan air outbreaks over the tropical North Atlantic. Pure Appl Geophys 119:677–691

Prospero JM, Lamb PJ (2003) African droughts and dust transport in the Caribbean: climate change implications. Science 302:1024–1027

Prospero JM, Nees RT (1977) Dust concentrations in the atmosphere of the equatorial North Atlantic: possible relationship to the Sahelian drought. Science 196:1196–1198

Prospero JM, Savoie DL (1989) Effect of continental sources of nitrate concentrations over the Pacific Ocean. Nature 339:678–689

Prospero JM, Bonatti E, Schuber C, Carlson TN (1970) Dust in the Caribbean traced to an African dust storm. Earth Planet Sci Lett 9:287–293

Prospero JM, Glaccum RA, Nees RT (1981) Atmospheric transport of soil dust from Africa to South America. Nature 289:570–572

Prospero JM, Nees RT, Uematsu M (1987) Deposition rate of particulate and dissolved aluminium derived from Saharan dust in precipitation at Miami, Florida. J Geophys Res 92:14723–14731

Prospero JM, Ginoux P, Torres O, Nicholson SE, Gill TE (2002) Environmental characterisation of global sources of atmospheric soil dust identified with the Nimbus 7 total ozone mapping spectrometer (TOMS) absorbing aerosol product. Rev Geophys 40:2–31

Psenner R (1999) Living in a dusty world: airborne dust as a key factor for alpine lakes. Water Air Soil Pollut 112:217–227

Pye K (1987) Aeolian dust and dust deposits. Academic Press, London

Pye K (1992) Aeolian dust transport and deposition over Crete and adjacent parts of the Mediterranean Sea. Earth Surf Process Landforms 17:271–288

Pye K, Sherwin D (1999) Loess. In: Goudie AS, Livingstone I, Stokes S (eds) Aeolian environments, sediments and landforms. Wiley, Chichester, pp. 214–238

Pye K, Sperling CHB (1983) Experimental investigation of silt formation by static breakage processes: the effect of temperature, moisture and salt on quartz dune sand and granitic regolith. Sedimentology 30:49–62

Pye K, Zhou L-P (1989) Late Pleistocene and Holocene Aeolian dust deposition in North China and the northwest Pacific Ocean. Palaeogeogr Palaeoclimatol, Palaeoecol 73:11–23

Pye K, Winspear NR, Zhou LP (1995)Thermoluminescence ages of loess and associated sediments in central Nebraska, USA. Palaeogeogr Palaeoclimatol Palaeoecol 118:73–87

Qian W, Quan L, Shi S (2002) Variations of the dust storm in China and its climatic control. J Clim 15:1216–1229

Qiang XK, Li Z, Li X, Powell C, Zheng HB (2001) Magnetostratigraphic record of the Late Miocene onset of the east Asian monsoon, and Pliocene uplift of northern Tibet. Earth Planet Sci Lett 187:83–93

Querol X, et al. (2004) Levels of particulate matter in rural, urban and industrial sites in Spain. Sci Total Environ 334/335:359–376

Quijano AL, Sokolik IN, Toon BO (2000) Radiative heating rates and direct radiative forcing by mineral dust in cloudy atmospheric conditions. J Geophys Res 105(D10):12207–12219

Rahn KA, Borys RA, Shaw GE (1977) The Asian source of Arctic haze bands. Nature 268:712–714

Rahn KA, Borys RA, Shaw GE (1981) Asian desert dust over Alaska: anatomy of an Arctic haze episode. In: Péwé TL (ed) Desert dust. Geol Soc Am Spec Pap 186: 37–70

Rajkumar WS, Chang AS (2000) Suspended particulate matter concentrations along the east–west corridor, Trinidad, West Indies. Atmos Environ 34:1181–1187

Raloff J (2001) Ill winds. Sci News 160:218–220

Rampach MR, Lu H (2004) Representation of land-surface processes in Aeolian transport models. Environ Model Software 19:93–112

Ramsperger B, Peinemann N, Stahr K (1998) Deposition rates and characteristics of Aeolian dust in the semi-arid and sub-humid regions of the Argentinean Pampa. J Arid Environ 39:467–476

Rao YP (1981)The climate of the Indian subcontinent. In: Takiatash K, Arakawa H (eds) Climates of Southern and Western Asia. (World survey of climatology, vol 9). Elsevier, Amsterdam, pp. 67–118

Rapp A (1984) Are terra rossa soils in Europe eolian deposits from Africa? Geol Foeren Stockholm Forh 105:161–168

Ratmeyer V, Balzer W, Bergametti G, Chiapello I, Fischer G, Wyputta U (1991) Seasonal impact of mineral dust on deep-ocean particle flux in the eastern Atlantic Ocean. Mar Geol 159:241–252

Ravi S, D'Odorico P, Over TM, Zobeck TM (2004) On the effect of air humidity on soil susceptibility to wind erosion: the case of air-dry soils. Geophys Res Lett 31:L09501

Rea DK (1989) Geologic record of atmospheric circulation on tectonic time scales. In: Leinen M, Sarnthein M (eds) Palaeoclimatology and palaeometeorology: modern and past patterns of global atmospheric transport. Kluwer, Dordrecht, pp. 842–857

Rea DK (1994)The paleoclimatic record provided by eolian deposition in the deep sea: the geologic history of wind. Rev Geophys 32:159–195

Rea DK, Leinen M (1988) Asian aridity and the zonal westerlies: late Pleistocene and Holocene record of eolian deposition in the northwest Pacific Ocean. Palaeogeogr Palaeoclimatol Palaeoecol 66:1–8

Read PL, Lewis SR (2004) The Martian climate revisited. Atmosphere and environment of a desert planet. Springer, Berlin Heidelberg New York

Reheis MC (1990) Influence of climate and eolian dust on the major-element chemistry and clay mineralogy of soils in the northern Bighorn Basin, USA. Catena 17:219–248

Reheis M (1997) Dust deposition of Owens (dry) Lake, 1991–1994: preliminary findings. J Geophys Res 102:25999–26008

Reheis M, Kihl R (1995) Dust deposition in southern Nevada and California, 1984–1989: relations to climate, source area and source lithology. J Geophys Res 100(D5):8893–8918

Reheis MC, Goodmacher JC, Harden JW (1995) Quaternary soils and dust deposition in southern Nevada and California. Bull Geol Soc Am 107:1003–1022

Reheis M, Budahn J, Lamothe P (1999) Elemental analysis of modern dust in Southern Nevada and Southern California. US Geological Survey Open File Report 99-0531. US Geological Survey, Washington, D.C.

Reid JS, et al. (2003) Analysis of measurements of Saharan dust by airborne and ground-based remote sensing methods during the Puerto Rico dust experiment (PRIDE). J Geophys Res 108 (D19), doi:10.1029/2002JD0022493

Reiff J, Forbes GS, Spieksma FTM, Reynders JJ (1986) African dust reaching north-western Europe: a case study to verify trajectory calculations. J Clim Appl Meteorol 25:1543–1567

Rendell HM (1984) New perspectives on the Pleistocene and Holocene sequences of the Potwar plateau and adjacent areas of northern Pakistan. In: Miller KJ (ed) The international Karakoram project, vol 1. Cambridge University Press, Cambridge, pp. 389–398

Rendell HM (1989) Loess deposition during the Late Plesitocene in northern Pakistan. Z Geomorphol Suppl 76:247–255

Resane T, Annegarn H, Freiman T (2004) The day of the white rain: origin of unusual dust deposition in Johannesburg, South Africa. S Afr J Sci 100:483–487

Rex RW, Syers JK, Jackson ML, Clayton RN (1969) Aeolian origin of quartz grains in soils of Hawaiian Islands and in Pacific pelagic sediments. Science 163:277–279

Reynolds R, Belnap J, Reheis M, Lamothe P, Luiszer F (2001) Aeolian dust in Colorado Plateau soils: Nutrient inputs and recent change in source. Proc Natl Acad Sci USA 98:7123–7127

Reynolds R, et al. (2003) Sources, transport paths, and impacts of atmospheric dust in the American southwest. Abstr Workshop Mineral Dust 2:10–12

Reyss JL, Pirazzoli PA, Haghipor A, Hatte C, Fontugne M (1998) Quaternary marine terraces and tectonic uplift rates on the south coast of Iran. Geol Soc Spec Publ 146:225–237

Richardson JSW (1981) Pink dustfall. J Meteorol 6:89–90

Richthofen F von (1882) On the mode of origin of the loess. Geol Mag 9:293–305

Ridgwell AJ (2002) Dust in the Earth system: the biogeochemical linking of land, air and sea. Philos Trans R Soc 360A:2905–2924

Ridgwell AJ (2003) Implications of the glacial CO_2 'iron hypothesis' for Quaternary climate change. Geochem Geophys Geosyst 4:1076

Ridgwell AJ, Watson AJ (2002) Feedback between Aeolian dust, climate and atmospheric CO_2 in glacial time. Paleoceanography 17, doi:10.1029/2001 PA000729

Riksen M, Spaan W, Arrué JL, López MV (2003a) What to do about wind erosion. In: Warren A (ed) Wind erosion on agricultural land in Europe. European Commission, Luxembourg, pp. 39–52

Riksen M, Brouwer F, Graaff J de (2003b) Soil conservation policy measures to control wind erosion in northwestern Europe. Catena 52:309–326

Riseborough RW, Huggett RJ, Griffin JJ, Goldberg ED (1968) Pesticide trans-Atlantic movements in the north–east trades. Science 159:1233–1236

Riser J (1985) Le rôle du vent au cours des derniers millénaires dans le Bassin Saharien d'Araouane (Mali). Bull Assoc Geogr Fr 62:311–317

Risser J (1985) Soil erosion problems in the USA. Desert Control Bull 12:20–25

Roberts HM, Muhs DR, Wintle AG, Duller GAT, Bettis EA (2003) Unprecedented last-glacial mass accumulation rates determined by luminescence dating of loess from western Nebraska. Quat Res 59:411–419

Robinson WO (1936) Composition and origin of dust in the fall of brown snow, New Hampshire and Vermont, February 24th, 1936. Mon Weather Rev 64:86

Roda F, Bellot J, Avila A, Escarre A, Pinol J, Terradas J (1993) Saharan dust and the atmospheric inputs of elements and alkalinity to Mediterranean ecosystems. Water Air Soil Pollut 66:277–288

Rodriguez S, Querol X, Alastney A, Kallos G, Kakaliagou O (2001) Saharan dust contributions to PM10 and TSP levels is southern and eastern Spain. Atmos Environ 35:2433–2447

Rögner K, Symkatz-Kloss W (1991) The deposition of eolian sediments in lacustrine and fluvial environments of Central Sinai (Egypt). Catena Suppl 20:75–91

Rognon P, Coudé-Gaussen G, Revel M, Grousset FE, Pédemay P (1996) Holocene Saharan dust deposition on the Cape Verde islands: sedimentological and Nd-Sr isotopic arguments. Sedimentology 43:359–366

Rogora M, Mosello R, Marchetto A (2004) Long-term trends in the chemistry of atmospheric deposition in northwestern Italy: the role of increasing Saharan dust deposition. Tellus 56B:426–434

Rokosh D, Bush ABG, Rutter NW, Ding Z, Sun J (2003) Hydrologic and geologic factors that influenced spatial variations in loess deposition in China during the last interglacial–glacial cycle: results from proxy climate and GCM analyses. Palaeogeogr Palaeoclimatol Palaeoecol 193:249–260

Rosenfeld D, Rudich Y, Lahav R (2001) Desert dust suppressing precipitation: a possible desertification feedback loop. Proc Natl Acad Sci USA 98:5975–5980

Rosenfield JE, Considine DB, Meade PE (1997) Stratospheric effects of Mount Pinatubo aerosol studied with a coupled two-dimensional model. J Geophys Res 102(D3):3649–3670

Rösner U (1989) "Loss" am Rande der Wustensteppe? Erdkunde 43:233–242

Rost KT (1997) Observations on the distribution and age of loess-like sediments in the high-mountain ranges of central Asia. Z Geomorphol Suppl 111:117–129

Röthlisberger R, Bigler M, Wolff EW, Joos F, Monnin E, Hutterli MA (2004) Ice core evidence of the extent of past atmospheric CO_2 change due to iron fertilisation. Geophys Res Lett 31:L16207

Rouchy JM, Servant M, Fournier M, Causse C (1996) Extensive carbonate algal bioherms in upper Pleistocene saline lakes of the central Altiplano of Bolivia. Sedimentology 43:973–993

Rousseau DD, et al. (2002) Abrupt millennial climatic changes from Nussloch (Germnay) Upper Weichselian eolian records during the Last Glaciation. Quat Sci Rev 21:1577–1582

Ruddiman WF (1997) Tropical Atlantic terrigenous fluxes since 25,000 years BP. Mar Geol 136:189–207

Russell RJ, Russell RD (1934) Dust storm of April 12, 1934, Baton Rouge, La. Mon Weather Rev 62:162–163

Rutherford S, Clark E, McTainsh GH, Simpson R, Mitchell C (1999) Characteristics of rural dust events shown to impact on asthma severity in Brisbane, Australia. Int J Biometeorol 42:217–225

Rutter NW, Rokosh D, Evans ME, Little EC, Chlachula J, Velichko A (2003) Correlation and interpretation of paleosols and loess across European Russia and Asia over the last interglacial–glacial cycle. Quat Res 60:101–109

Safar MI (1985) Dust and dust storms in Kuwait. State of Kuwait, Kuwait

Saint-Amand P, Mathews LA, Gaines C, Reinking R (1986) Dust storms from Owens and Mono valleys, California. (Report TP 6731). Naval Weapons Center, China Lake, 79 pp

Sala JQ, Cantos JO, Chiva EM (1996) Red dust within the Spanish Mediterranean area. Clim Change 32:215–228

Sañudo-Wilhelmy SA, Flegal AR (2003) Potential influence of Saharan dust on the chemical composition of the Southern Ocean. Geochem Geophys Geosyst 4, doi:10.1029/2003GC000507

Sapozhnikova SA (1973) Map diagram of the number of days with dust storms in the hot zone of the USSR and adjacent territories, Report HT-23-0027. US Army Foreign and Technology Center, Charlottesville

SARH (1985) Proyecto Texcoco. Comisión del Lago de Texcoco, Mexico City, 16 pp

Sarnthein M, Koopmann B (1980) Late Quaternary deep-sea record on northwest African dust supply and wind circulation. Palaeoecol Afr 12:239–253

Sarnthein M, Thiede J, Pflaumann U, Erlenkeuser H, Futterer D, Koopman B, Lange H, Seibold E (1982) Atmospheric and oceanic circulation patterns off northwest Africa during the past 25 million years. In: Rad U von, Hinz K, Sarnthein M, Seibold E (eds) Geology of the northwest African continental margin. Springer, Berlin Heidelberg New York, pp. 545–604

Sarthou G, et al. (2003) Atmospheric iron deposition and sea-surface dissolved iron concentrations in the eastern Atlantic Ocean. Deep Sea Res I 50: 1339–1352

Sartori M, Evans ME, Heller F, Tsatskin A, Han JM (2005) The last glacial/interglacial cycle at two sites in the Chinese Loess Plateau: mineral magnetic, grain-size and [10]Be measurements and estimates of palaeoprecipitation. Palaeogeogr Palaeoclimatol Palaeoecol 222:145–160

Sassen K (2005) Dusty ice clouds over Alaska. Nature 434:456

Sassen K, DeMott PJ, Prospero JM (2003) Saharan dust storms and indirect effects on clouds: CRYSTAL-FACE results. Geophys Res Lett 30:1633

Savoie DL, Prospero JM (1982) Particle size distribution of nitrate and sulphate in the marine atmosphere. Geophys Res Lett 9:1207–1210

Sayago JM, Collantes MM, Karlson A, Sanabria J (2001) Genesis and distribution of the Late Pleistocene and Holocene loess of Argentina: a regional approximation. Quat Int 76/77:247–257

Schäfer W (1991) Quantifizizung der Bodenerosion durch Winds: Vorstellung eines BMFT-Forschungs projects. Mitt Dtsch Bodenkund Ges 65:47–50

Schlatter T (1995) Long distance dust.Weatherwise 48:38–39

Schlesinger WH (1985) The formation of caliche in soils of the Mojave Desert, California. Geochem Cosmochim Acta 49:57–66

Schollaert SE, Merrill JT (1998) Cooler sea surface west of the Sahara Desert correlated to dust events. Geophys Res Lett 25:3529–3532

Schramm CT (1989) Cenozoic climatic variation recorded by quartz and clay minerals in North Pacific sediments. In: Leinen M, Sarnthein M (eds) Palaeoclimatology and palaeometeorology: modern and past patterns of global atmospheric transport, Kluwer, Dordrecht, pp. 805–839

Schroeder JH (1985) Eolian dust in the coastal desert of the Sudan: aggregates cemented by evaporites. J Afr Earth Sci 3:370–380

Schubert SD, Suarez MJ, Pegion PJ, Koster RD, Bachmeister JT (2004) On the cause of the 1930s Dust Bowl. Science 303:1855–1859

Schüssler U, BalzerW, Deeken A (2005) Dissolved Al distribution, particulate Al fluxes and coupling to atmospheric Al and dust deposition in the Arabian Sea. Deep Sea Res II 52:1862–1878

Schütz L (1979) Saharan dust transport over the North Atlantic Ocean – model calculations and measurements. In: Morales C (ed) Saharan dust. Wiley, Chichester, pp. 267–278

Schütz L (1980) Long range transport of desert dust with special emphasis on the Sahara. Ann NY Acad Sci 338:515–532

Schütz L (1987) Atmospheric mineral dust – properties and source markers. In: Leinen M, Sarnthein M (eds) Palaeoclimatology and palaeometeorology: modern and past patterns of global atmospheric transport. Kluwer, Dordrecht, pp. 359–383

Schütz L, Jaenicke R, Pietrek H (1981) Saharan dust transport over the North Atlantic Ocean. In: Péwé T L (ed) Desert dust. Geol Soc Am Spec Pap 186:87–100

Schwab GO, Frevert RK, Edminster TW, Barnes KK (1966) Soil and water conservation engineering. Wiley, Chichester

Schwikowski M, Seibert P, Baltensperger U, Gäggeler HW (1995) A study of an outstanding Saharan dust event at the high-alpine site Jungfraujoch, Switzerland. Atmos Environ 29:1828–1842

Seppälä M (2004) Wind as a geomorphic agent in cold climates. Cambridge University Press, Cambridge

Shao Y (2000) Physics and modelling of wind erosion. Kluwer, Dordrecht

Shao Y, Wang JJ (2003) A climatology of Northeast Asian dust events. Meteorol Z 12:175–183

Shao Y, et al. (2003) Northeast Asia dust storms: real-time numerical prediction and validation. J Geophys Res 108 (22D):4691, doi: 10.1029/2003JD003667

Shaw GE (1980) Transport of Asian desert aerosol to the Hawaiian Islands. J Appl Meteorol 19:1254–1259

Shi P, Yan P, Yuan Y, Nearing MA (2004) Wind erosion research in China: past, present and future. Progr Phys Geogr 28:366–386

Shichang K, Dahe1 Q, Mayewski P A, Wake C P, Jiawen R (2001) Climatic and environmental records from the Far East Rongbuk ice core, Mt. Qomolangma (Mt. Everest). Episodes 24:176–181

Shikula NK (1981) Prediction of dust storms from meteorological observations in the South Ukraine, USSR. In: Péwé TL (ed) Desert dust. Geol Soc Am Spec Pap 186:261–266

Shinn EA, Smith GW, Prospero JM, Betzer P, Hayes ML, Garrison V, Barber RT (2000) African dust and the demise of Caribbean coral reefs. Geophys Res Lett 27:3029

Singer A, Ganor E, Dultz S, Fischer W (2003) Dust deposition over the Dead Sea. J Arid Environ 53:41–59

Singer MJ, Shainberg I (2004) Mineral soil surface crusts and wind and water erosion. Earth Surf Process Landforms 29:1065–1075

Sirocko F (1991) Deep-sea sediments of the Arabian Sea: a palaeoclimatic record of the southwest-Asian summer monsoon. Geol Rundsch 80:557–566

Sirocko F, Sarnthein M, Large H, Erlenkeuser H (1991) Atmospheric summer circulation and coastal upwelling in the Arabian Sea during the Holocene and the Last Glaciation. Quat Res 36:72–93

Sivall T (1957) Sirocco in the Levant. Geogr Ann 39:114–142

Skidmore EL (1986) Wind erosion control. Clim Change 9:209–218

Slate JL, Bull WB, Ku T-L, Shafiqullah M, Lynch DJ, Huang Y-P (1991) Soil-carbonate genesis in the Pinacate volcanic field, northwestern Sonora, Mexico. Quat Res 35:400–416

Smalley IJ (1966) The properties of glacial loess and the formation of loess deposits. J Sediment Petrol 36:669–676

Smalley IJ (ed) (1975) Loess. Lithology and genesis. Dowden, Hutchinson and Ross, Stroudsburg, Penn.

Smalley IJ, Krinsley DH (1978) Loess deposits associated with deserts. Catena 5:53–66

Smalley IJ, Vita-Finzi C (1968) The formation of fine particles in sandy deserts and the nature of 'desert' loess. J Sediment Petrol 38:766–774

Smalley IJ, Kumar R, O'Hara Dhand K, Jefferson IE, Evans RD (2005). The formation of silt material for terrestrial sediments: particularly loess and dust. Sediment Geol 179:321–328

Smirnov A, Holben BN, Dobovik O, O'Neill NT, Eck TF, Westphal DL, Goroch AK, Pietras C, Slutsker I (2002) Atmospheric aerosol optical properties in the Persian Gulf. J Atmos Sci 59:620–634

Smith BJ, McGreevy JP, Whalley WB (1987) The production of silt-size quartz by experimental salt weathering of a sandstone. J Arid Environ 12:199–214

Smith BJ, Wright JS, Whalley WB (2002) Sources of non-glacial, loess-size quartz silt and the origins of "desert loess". Earth Sci Rev 59:1–26

Smith GT, Ives LD, Nagelkerken IA, Ritchie K (1996) Caribbean sea fan mortalities. Nature 383:487

Smith J (1988) The dust fall of 27 October 1987 over Britain. J Meteorol 13:115–120

Smith J, Vance D, Kemp RA, Archer C, Toms P, King M, Zárate M (2003) Isotopic constraints on the source of Argentinian loess – with implications for atmospheric circulation and the provenance of Antarctic dust during recent glacial maxima. Earth Planet Sci Lett 212:181–196

Smith RM, Twiss PC, Krauss RK, Brown MJ (1970) Dust deposition in relation to site, season and climatic variables. Proc Soil Sci Soc Am 34:112–117

Soreghan GS, Soreghan MJ (2002) Atmospheric dust ansd algal dominance in the Late Paleozoic: a hypothesis. J Sediment Res 72:457–461

Soreghan GS, Elmore RD, Lewchuk MT (2002) Sedimentological-magnetic record of western Pangean climate in upper Paleozoic loessite (lower Cutler beds, Utah). Bull Geol Soc Am 114:1019–1035

Sprigg RC (1982)Alternating wind cycles of the Quaternary era and their influence on Aeolian sedimentation in and around the dune deserts of south-eastern Australia. In: Wasson RJ (ed) Quaternary dust mantles of China, New Zealand and Australia. (Proceedings of a workshop at the Australian National University, Canberra, 3–5 December 1980). Australian National University, Canberra, pp. 211–240

Stabell B (1989) Deflation and humidity during the past 700,000 years in NW Africa from the marine record. In: Leinen M, Sarnthein M (eds) Palaeoclimatology and palaeometeorology: modern and past patterns of global atmospheric transport. Kluwer, Dordrecht, pp. 771–777

State of Oregan (2004) Emergency management plan, natural hazards mitigation plan, DS1–DS13. State of Oregan, Salem

State of the Environment Advisory Council (1996) Australian state of the environment. CSIRO, Collingwood

Stein R (1985) The post-Eocene sediment record of DSDP site 366: implications for African climate and plate tectonic drift. Geol Soc Am Mem 163:305–315

Steinbeck J (1939) The grapes of wrath. Viking Press, New York

Stensland GJ, Semorin RG (1982) Another interpretation of the pH trend in the United States. Bull Am Meteorol Soc 63:1277–1284

Sterk G (2003) Causes, consequences and control of wind erosion in sahelian Africa: a review. Land Degrad Dev 14:95–108

Stetler LD, Gaylord DR (1996) Evaluating eolian–climatic interactions using a regional climate model from Hanford, Washington USA. Geomorphology 17:99–113

Stevenson CM (1969) The dust fall and severe storms of 1 July 1968. Weather 24:126–132

Stewart RA, Pilkey OH, Nelson PW (1965) Sediments of the northern Arabian Sea. Mar Geol 3:411–427

Stout JE (2001) Dust and environment in the southern High Plains of North America. J Arid Environ 47:425–441

Stout JE, Lee JA (2003) Indirect evidence of wind erosion trends on the southern High Plains of North America. J Arid Environ 55:43–61

Strausberg MJ, Wang H, Richardson MI, Ewald SP, Toigo AD (2005) Observations of the initiation and evolution of the 2001 Mars global dust storm. J Geophys Res 110:E02006, doi:10.1029/2004JE002361

Sturman A, Tapper N (1996) The weather and climate of Australia and New Zealand. Oxford University Press, Oxford

Stuut J-B, Zabel M, Ratmeter V, Helmke P, Schefuss E, Lavik G, Schneider R (2005) Provenance of present-day eolian dust collected off NW Africa. J Geophys Res 110, doi:10.1029/2004JD005161

Subba Row DV, Al-Yamani F, Nageswara Rao CV (1999) Eolian dust effects phytoplankton in the waters off Kuwait, the Arabian Gulf. Naturwissenschaften 86:525–529

Sugden W (1963) Some aspects of sedimentation in the Persian Gulf. J Sediment Petrol 33:355–364

Sultan B, Labadi K, Guégan J, Janicot S (2005) Climate drives the meningitis epidemics onset in west Africa. PLoS Med 2:e6

Summerfield MA (1983) Silcrete. In: Goudie AS, Pye K (eds) Chemical sediments and geomorphology. Academic Press, London, pp. 59–91

Sun J (2000) Origin of sand mobilization during the past 2300 years in the Mu Us Desert, China. Quat Res 53:78–88

Sun J (2002a) Provenance of loess material and formation of loess deposits on the Chinese Loess Plateau. Earth Planet Sci Lett 203:845–859

Sun J (2002b) Source regions and formation of the loess sediments on the high mountain regions of northwestern China. Quat Res 58:341–351

Sun J (2005) Nd and Sr isotopic variations in Chinese eolian deposits during the past 8 Ma: implications for provenance change. Earth Planet Sci Lett 240:454–466

Sun D, Shaw J, Zhisheng A, Minyang G, Leping Y (1998) Magnetostratigraphy and paleoclimatic interpretation of continuous 7.2 Ma Late Cenozoic eolian sediments from the Chinese Loess Plateau. Geophys Res Lett 25:85–88

Sun D, Bloemendal J, Rea DK, An Z, Vandenberghe J, Lu H, Su R, Liu T (2004) Biomodal graiz-size distribution of Chinese loess and its palaeoclimatic significance. Catena 55:325–340

Svennson A, Biscaye PE, Grousset FE (2000) Characterization of late glacial continental dust in the Greenland Ice Core Project ice core. J Geophys Res 105 (D4):4637–4656

Swap R, Garstang M, Greco S, Talbot R, Kallberg P (1992) Saharan dust in the Amazon Basin. Tellus 44B:133–149

Swap R, Ulanski S, Cobbett M, Garstang M (1996) Temporal and spatial characteristics of Saharan dust outbreaks. J Geophys Res 101(D2):4205–4220

Sweeney M (1998) Saharan dust fall in western Ireland. J Meteorol 23:140

Syers JK, Jackson ML, Berkeisev VE, Clayton RN, Rex RW (1969) Eolian sediment influence on pedogenesis during the Quaternary. Soil Sci 107:421–427

Syers JK, Mokma DL, Jackson ML, Dolcater DL, Rex RW (1972) Mineralogical composition and caesium-137 retention properties of continental aerosolic dusts. Soil Sci 113:116–121

Ta W, Xiao H, Qu J, Xiao Z, Yang G, Wang T, Zhang X (2004a) Measurements of dust deposition in Gansu Province, China, 1986–2000. Geomorphology 57:41–51

Ta W, Wang T, Xiao H, Zhu X, Xiao Z (2004b) Gaseous and particulate air pollution in the Lanzhou Valley, China. Sci Total Environ 320:163–176

Takemi T, Seino N (2005) Dust storms and cyclone tracks over the arid regions in east Asia in spring. J Geophys Res 110 (D), doi:1029/2004JD004698

Takemura T, Uno I, Nakajima T, Higurashi A, Sano I (2002) Modeling study of long-range transport of Asian dust and anthropogenic aerosols from east Asia. Geophys Res Lett 29, doi: 10.1029/2002GL016251

Talbot RW, Harriss RC, Browell EV, Gregory GL, Sebacher DI, Beck SM (1986) Distribution and geochemistry of aerosols in the tropical North Atlantic troposhere: relationship to Saharan dust. J Geophys Res 91(D4):5173–5182

Talbot RW, Dibb JE, Lefer BL, Bradshaw JD, Sandholm ST, Blake DR, Blake NJ, Sachse GW, Collins JE, Heikes BG, Merrill JT, Gregory GL, Anderson BE, Singh HB, Thornton DC, Bandy AR, Purschel RF (1997) Chemical characteristics of continental outflow from Asia to the troposphere over the western Pacific Ocean during February–March 1994: Results from PEM-West B. J Geophys Res 102:28255–28274

Tanaka TY, Kurosaki Y, Chiba M, Matsumura T, Nagai T, Yamazaki A, Uchiyama A, Tsunematsu N, Kai K (2005) Possible transcontinental dust transport from North Africa and the Middle East to East Asia. Atmos Environ 39:3901–3909

Tegen I (2003) Modelling the mineral dust aerosol in the climate system. Quat Sci Rev 22:1821–1834

Tegen I, Fung I (1994) Modelling of mineral dust in the atmosphere: sources, transport and optical thickness. J Geophys Res 99(D11):22879–22914

Tegen I, Fung I (1995) Contribution to the atmospheric mineral aerosol load from land surface modification. J Geophys Res 100(D9):18707–18726

Tegen I, Lacis AA, Fung I (1996)The influence of climate forcing of mineral aerosols from disturbed soils. Nature 380:419–422

Tegen I, Werner M, Harrison SP, Kohfled KE (2004) Relative importance of climate and land use in determining present and future global soil dust emissions. Geophys Sci Rev 2004:L05105

Teruggi ME (1957) The nature and origin of Argentine loess. J Sediment Petrol 27:322–332

Tetzlaff G, Peters M (19860 Deep-sea sediments in the eastern equatorial Atlantic off the African coast and meteorological flow patterns over the Sahel. Geol Rundsch 75:71–79

Tetzlaff G, Peters M, Janssen W, Adams LJ (1989) Aeolian dust transport in West Africa. In: Leinen M, Sarnthein M (eds) Palaeoclimatology and palaeometeorology: modern and past patterns of global atmospheric transport. Kluwer, Dordrecht, pp. 1985–2003

Thiagarajan N, Lee CTA (2004) Trace-element evidence for the origin of desert varnish by direct aqueous atmospheric deposition. Earth Planet Sci Lett 224:131–141

Thiede J (1979) Wind regimes over the late Quaternary southwest Pacific Ocean. Geology 7:259–262

Thiel E (1944) Staubstürme in Südostrussland. Petermann Geogr Mitt 90:238

Thomas DSG, Shaw P (1991) The Kalahari environment. Cambridge University Press, Cambridge

Thomas DSG, Knight M, Wiggs GFS (2005) Remobilization of southern African desert dune systems by twenty-first century global warming. Nature 435:1218–1221

Thomas FG (1982) Saharan dust-fall in Dover, Kent. J Meteorol 7:92–93

Thomas FG (1983) Dust fall-out over Kent and Sussex. J Meteorol 8:126–127

Thomas FG (1985) Weather of 8 November 1984 and the Saharan dust-fall. J Meteorol 10:147–148

Thomas FG (1993) Dust rain.Weather 48:193

Thomas PC, Gierasch P, Sullivan R, Miller DS, Alvarez del Castillo E, Cantor B, Mellon MT (2003) Mesoscale linear streaks on Mars: environments of dust entrainment. Icarus 162:242–258

Thompson LG, Mosley-Thompson E (1981) Microparticle concentration variations linked with climatic changes: evidence from polar ice cores. Science 212:812–815

Thompson LG, Mosley-Thompson E, Davis JF, Bolzan JF Dai J, Klei L (1990) Glacial stage ice core records from the subtropical Dunde Ice Cap, China, Ann Glaciol 14:288–297

Thompson LG, Mosley-Thompson E, Davis ME, Lin P-N, Henderson KA, Cole-Dai J, Bolzan JF, Liu, K-B (1995) Late Glacial Stage and Holocene tropical ice core records from Huascarán, Peru. Science 269:46–50

Thompson LG, et al. (1998) A 25,000-year tropical climate history from Bolivian ice cores. Science 282:1858–1864

Thornthwaite CW (1948) An approach towards a rational classification of climate. Geogr Rev 38:55–94

Tibke G (1988) Basic principles of wind erosion control. Agric Ecosyst Environ 22/23:103–122

Tiessen H, Hauffe H-K, Mermut AR (1991) Deposition of Harmattan dust and its influence on base saturation of soils in northern Ghana. Geoderma 49:285–299

Tiller KG, Smith LH, Merry RH (1987) Accessions of atmospheric dust east of Adelaide, South-Australia, and the implications for pedogenesis. Aust J Soil Res 25:43–54

Tindale NW, Pease PP (1999) Aerosols over the Arabian Sea: atmospheric transport pathways and concentrations of dust and sea salt. Deep Sea Res II 46:1577–1595

Todhunter PE, Cihacek LJ (1999) Historical reduction of airborne dust in the Red River Valley of the North. J Soil Water Conserv 54:543–551

Tomadin L (1974) Les minéraux argileux dans les sediments actuels de la Mer Tyrrhénienne. Bull Groupe Fr Argiles 26:219–228

Tomadin L, Lenaz R (1989) Eolian dust over the Mediterranean and their contribution to the present sedimentation. In: Leinen M, Sarnthein M (eds) Palaeoclimatology and palaeometeorology: modern and past patterns of global atmospheric transport. (NATO ASI Series C 282). Kluwer, Dordrecht, pp. 267–282

Tomadin L, Lenaz R, Landuzzi V, Mazzucotelli A, Vannucci R (1984) Wind-blown dusts over the central Mediterranean. Oceanol Acta 7:13–23

Toon OB (2003) African dust in Florida clouds. Nature 424:623–624

Torii M, Lee T-Q, Fukuma K, Mishima T, Yamazaki T, Oda H, Ishikawa N (2001) Mineral magnetic study of the Taklimakan desert sands and its relevance to the Chinese loess. Geophys Res Int 146:416–424

Torres O, Bhartia PK, Herman JR, Ahmad Z, Gleason J (1998) Derivation of aerosol properties from satellite measurements of backscattered ultraviolet radiation: theoretical basis. J Geophys Res 103:17099–17110

Torres-Padrón ME, Gelado-Caballero MD, Collado-Sánchez C, Siruela-Matos VF, Cardona-Castellano PJ, Hernández-Brito JJ (2002) Variability of dust inputs to the CANIGO zone. Deep Sea Res II 49:3455–3464

Tothill JD (1948) Agriculture in the Sudan. Oxford University Press, Oxford

Tout DG, Kemp V (1985) The named winds of Spain. Weather 40:322–329

Toyoda S, Naruse T (2002) Eolian dust from the Asian deserts to the Japanese Islands since the Last Glacial Maximum: the basis for the ESR method. Trans Jpn Geomorphol Union 23/25:811–820

Tricart J, Cailleux A (1969) Le modèle des régions sèches (2 vols). CDU, Paris

Trilsbach A, Hulme M (1984) Recent rainfall changes in central Sudan and their physical and human implications. Trans Inst Br Geogr 9:280–298

Tsoar H, Pye K (1987) Dust transport and the question of desert loess formation. Sedimentology 34:139–153

Tsunematsu N, Sato T, Kimura F, Kai K, Kurosaki Y, Nagai T, Zhou HF, Mikami M (2005) Extensive dust outbreaks following the morning inversion breakup in the Taklimakan Desert. J Geophys Res 110D21: D21207

Tsyganenko AF (1968) Aeolian migration of water soluble material and its probable geochemical and soil formation significance. Trans Int Congr Soil Sci 9:333–341

Tucker CJ, Nicholson SE (1999) Variations in the size of the Sahara Desert from 1980 to 1997. Ambio 28:587–591

Tullet MT (1978) A dust-fall on 6 March 1977. Weather 33:48–52

Tullet MT (1980) A dust-fall of Saharan origin on 15 May 1979. Royal Dublin Society. J Earth Sci 3:35–39

Tullet MT (1988) A dust fall on 1 September 1987. Weather 43:340

Tullett MT (1984) Saharan dust-fall in Northern Ireland. Weather 39:151–152

Turekian KK, Cochran JK (1981) 210Pb in surface air at Eniwetok and the Asian dust flux in the Pacific. Nature 292:522–524

Twenhofel WH (1950) Principles of sedimentation. McGraw Hill, New York

Tyson PD, Preston-Whyte RA (2000)The atmosphere, weather and climate of southern Africa. Oxford University Press, Cape Town

UNEP (1992) Environmental issues in the Aral Sea basin. UN Environment Programme, Nairobi

Udden JA (1898) The mechanical composition of wind deposits. Augustana Library, Rock Island, Ill.

Urvoy Y (1942) Les bassins du Niger. Mem Inst Fond Afr Noire 4

VanCuren RA, Cahill TA (2002) Asian aerosols in North America frequency and concentration of fine dust. J Geophys Res 107, doi:10.1029/2002JD002204

Van Den Heever SC, Carrio GG, Cotton WR, Straka WC (2005) The impacts of Saharan dust on Florida storm characteristics. Abstr Am Meteorol Soc Annu Meet 2005:5.1

Van Heuklon TK (1977) Distant source of 1976 dustfall in Illinois and Pleistocene weather models. Geology 5:693–695

VDL (1902) La grande pluie de poussière de 9 au 12 Mars, 1901. Ciel Terre 23:318–321

Vidic NJ, Montañez IP (2004) Climatically driven glacial–interglacial variations in C3 and C4 plant proportions on the Chinese Loess Plateau. Geology 32:337–340

Vine H (1987)Wind-blown materials and W. African soils: an explanation of the 'ferralitic soil over loose sandy sediments' profile. Geol Soc Spec Pub 35:171–183

Vink S, Measures CI (2001) The role of dust deposition in determining surface water distributions of Al and Fe in the South West Atlantic. Deep Sea Res II 48: 2787–2809

Visser SM, Sterk G, Karssenberg D (2005) Wind erosion modelling in a Sahelian environment. Environ Model Software 20:69–84

Vita-Finzi C (1981)Late Quaternary deformation on the Makran coast of Iran. Z Geomorphol Suppl 40:213–226

Wagenbach D, Geis K (1989) The mineral dust record in a high alpine glacier (Colle Gnifett, Swiss Alps). In: Leinen M, Sarnthein M (eds) Paleoclimatology and paleometeorology: modern and past patterns of global atmospheric transport. Kluwer, Dordrecht, pp. 543–564

Wake C P, Mayewski PA, Li Z, Han J, Qin D (1994) Modern eolian dust deposition in central Asia. Tellus 46B:220–233

Walker PH, Costin AB (1971) Atmospheric dust accession in south-eastern Australia. Aust J Soil Res 9:1–5

Wang G, Tuo W, Du M (2004a) Flux and composition of wind-eroded dust from different landscapes of an arid inland river basin in north-western China. J Arid Environ 58:373–385

Wang S (2001) Fighting dust storms: the case study of Canada's prairie region. In: YangY, Squires V, Lu Q (eds) Global alarm: dust and sandstorms from the world's drylands. United Nations, Bangkok, pp. 77–107

Wang T, Dong G (1994) Sand sea history of the Taklimakan for the past 30 000 years. Geogr Ann 76A:131–141

Wang X, Dong Z, Zhang J, Liu L (2004b) Modern dust storms in China: an overview. J Arid Environ 58:559–574

Wang S, Wang J, Zhou Z, Shang K (2005a) Regional characteristics of three kinds of dust storm events in China. Atmos Environ 39:509–552

Wang YQ, Zhang XY, Arimoto R, Cao JJ, Shen ZX (2005b) Characteristics of carbonate content and carbon and oxygen isotopic composition of northern China soil and dust aerosol and its application to tracing dust sources. Atmos Environ 39: 2631–2642

Ward A (1950) Dust storm at North Front, Gibraltar. Meteorol Mag 80:196–198

Warn GF, Cox WH (1951) A sedimentological study of dust storms in the vicinity of Lubbock, Texas. Am J Sci 249:553–568

Warner TT (2004) Desert meteorology. Cambridge University Press, Cambridge

Washington R, Todd MC (2005) Atmospheric controls on mineral dust emission from the Bodélé Depression, Chad: the role of the low level jet. Geophys Res Lett 32, doi:10.1029/2005GL023597

Washington R, Todd M, Middleton N, Goudie AS (2003) Global dust storm source areas determined by the total ozone monitoring spectrometer and ground observations. Ann Assoc Am Geogr 93:297–313

Wasson RJ, Fitchett K, Mackey B, Hyde R (1988) Large-scale patterns of dune type, spacing and orientation in the Australian continental dunefield. Aust Geogr 19:89–104

Watson A (1979) Gypsum crusts in deserts. J Arid Environ 2:3–20

Watson EH (1934) Note on the dust storm of November 13, 1933. Science 79:320

Weaver JE, Flory EL (1937) Stability of climax prairie and some environmental changes resulting from breaking. Ecology 15:333–347

Weir-Brush JR, Garrison VH, Smith GW, Shinn EA (2004)The relationship between Gorgonian Coral (Cnidaria: Gorgonacea) diseases and African dust storms. Aerobiologia 20:119–126

Weiss D, Shotyk W, Rieley J, Page S, Gloor M, Reese S, Martinez-Cortizas A (2002) The geochemistry of major and selected trace elements in a forested peat bog, Kalimantan, SE Asia, and its implications for past atmospheric dust deposition. Geochim Cosmochim Acta 66:2307–2323

Wellington JH (1938) The Kunene River and the Etosha Plain. S Afr Geogr J 20:21–32

Wells GL, Zimbelman JR (1997) Extraterrestrial and surface processes. In: Thomas DSG (ed) Arid zone geomorphology: process, form and change in drylands, 2nd edn.Wiley, Chichester, pp. 659–690

Wells SG, McFadden LD, Dohrenwend JC (1987) Influence of late-Quaternary climatic changes on geomorphic and pedogenic processes on a desert piedmont. Quat Res 27:130–146

Wentworth CK (1922)A scale of grade and class terms for clastic sediments. J Geol 30:377–392

Werner M, Tegen I, Harrison SP, Kohfeld KE, Prenctice IC, Balkanski Y, Rodhe H, Roelandt C (2002) Seasonal and interannual variability of the mineral dust cycle under present and glacial climate conditions. J Geophys Res 107:D24, doi: 10.1029/2002JD002365

Westphal DL, Toon OB, Carlson TN (1987) A two dimensional numerical investigation of the dynamics and microphysics of Saharan dust storms. J Geophys Res 92D:3027–3049

Wheaton EE (1990) Frequency and severity of drought and dust storms. Can J Agric Econ 38:695–700

Wheaton EE (1992) Prairie dust storms – a neglected hazard. Nat Hazards 5:53–63

Wheaton EE, Chakravarti AK (1987) Some temporal, spatial and climatological aspects of dust storms in Saskatchewan. Climatol Bull 21:5–16

Wheaton EE, Chakravarti AK (1990) Dust storms in the Canadian prairies. Int J Climatol 10:829–837

Wheeler DA (1986) The meteorological background to the fall of Saharan dust, November 1984. Meteorol Mag 115:1–9

White BR (1979) Soil transport by wind on Mars. J Geophys Res 84:4643–4651

Whitney MI, Dietrich RV (1973) Ventifact sculpture by windblown dust. Bull Geol Soc Am 84:2561–2582

Wicks GA, Smith DE (1973) Chemical fallow in a winter wheat–fallow rotation. Weed Sci 21:97–102

Wiggs GFS, O'Hara S, Wegerdt J, Van Der Meer J, Small I, Hubbard R (2003) The dynamics and characteristics of aeolian dust in dryland Central Asia: possible impacts on human exposure and respiratory health in the Aral Sea basin. Geogr J 169:142–157

Wigner KA, Peters RE (1987) Synoptic climatologies of blowing dust on the Texas South Plains, 1947–1984. J Arid Environ 13:199–209

Wilke BM, Duke BJ, Jimoh WLO (1984) Mineralogy and chemistry of Harmattan dust in northern Nigeria. Catena 11:91–96

Williams GE (1970) Piedmont sedimentation and late Quaternary chronology in the Biskra region of the northern Sahara. Z Geomorphol Suppl 10:40–63

Williams MAJ, Clarke MF (1995) Quaternary geology and prehistoric environments in the Son and Belan valleys, north central India. Mem Geol Soc India 32:282–308

Williams P, Young M (1999) Costing dust. (CSIRO Land and Water, Policy and Economic Research Unit, Final Report). CSIRO, Canberra, 36 pp

Williams RBG (1975) The British climate during the last glaciation: an interpretation based on periglacial phenomena. In: Wright AE, Moseley F (eds) Ice ages: ancient and modern. Seel House Press, Liverpool, pp. 95–120

Willis DM, Easterbrook MG, Stephenson FR (1980) Seasonal variation of oriental sunspot sightings. Nature 287:617–619

Wilshire HG (1980) Human causes of accelerated wind erosion in California's deserts. In: Coates DR, Vitek JD (eds) Geomorphic thresholds. Dowden, Hutchinson & Ross, Stroudsburg, pp. 415–433

Wilshire HG, Nakata JK, Hallet B (1981) Field observations of the December 1977 wind storm, San Joaquin valley, California. In: Péwé TL (ed) Desert dust: origin, characteristics and effects on man. Geol Soc Am Spec Pap 186:233–251

Winchell AN, Miller ER (1918) The dustfall of March 1918. Am J Sci 46:559–609; 47:133–134

Windom HL (1969) Atmospheric dust records in permanent snowfields: implications to marine sedimentation. Bull Geol Soc Am 80:761–782

Winspear NR, Pye K (1995) Textural, chemical and mineralogical evidence for the origin of Peoria Loess in central and southern Nebraska, USA. Earth Surf Process Landforms 20:735–745

Wolcken K (1951) Descripcion de una violenta tempestad de polvo. Meteoros 1:211–216

Wong S, Desler AE (2005) Suppression of deep convection over the tropical North Atlantic by the Saharan Air Layer. Geophys Res Lett 32, doi:10.1029/2004GL022295

Woodruff NP (1956) Windblown soil abrasive injuries to winter wheat plants. J Agron 48:499–505

Woodruff NP, Siddoway FH (1965) A wind erosion equation. Proc Soil Sci Soc Am 29:602–608

Woodruff NP, Chepil WS, Lynch RD (1957) Emergency chiselling to control wind erosion. Kans Agric Exp Stn Tech Bull 90

Woodward NP (1897) The dry lakes of Western Australia. Geol Mag 4:363–366

Worster D (1979) Dust Bowl. The Southern Plains in the 1930s. Oxford University Press, New York

Wright J (2001a) Making loess-sized silt: data from laboratory simulations and implications for sediment transport pathways and the formation of 'desert' loess deposits associated with the Sahara. Quat Int 76/77:7–19

Wright J (2001b) "Desert" loess versus "glacial" loess: quartz silt formation, source areas and sediment pathways in the formation of loess deposits. Geomorphology 36:231–256

Wright J (2002) Granitoid weathering profiles as a source of loessic silt. Trans Jpn Geomorphol Union 23/25:769–793

Wu P-C, Tsai J-C, Li F-C, Lung S-C, Su H-J (2004) Increased levels of ambient fungal spores in Taiwan are associated with dust events from China. Atmos Environ 38:4879–4886

Xie S, Yu T, Zhang Y, Zeng L, Qi L, Tang X (2005) Characteristics of PM_{10}, SO_2, NO_x and O_3 in ambient air during the dust storm period in Beijing. Sci Total Environ 345:153–164

Xuan J, Liu G, Du K (2000) Dust emission inventory in Northern China. Atmos Environ 34:4565–4570

Xuan J, Sokolik IN, Hao J, Guo F, Mao H, Yang G (2004) Identification and characterisation of atmospheric mineral dust in East Asia. Atmos Environ 38:6239–6252

Yaalon DH (1987) Saharan dust and desert loess: effect on surrounding soils. J Afr Earth Sci 6:569–571

Yaalon DH, Dan J (1974) Accumulation and distribution of loess-derived deposits in the semi-desert and desert fringe of Israel. Z Geomorphol Suppl 20:91–105

Yaalon DH, Ganor E (1973) The influence of dust on soils during the Quaternary. Soil Sci 116:146–155

Yaalon, DH, Ganor E (1975) Rate of aeolian dust accretion in the Mediterranean and desert fringe environments of Israel. Int Conf Sediment 19:169–174

Yaalon DH, Ganor E (1979) East Mediterranean trajectories of dust-carrying storms from the Sahara and Sinai. In: Morales C (ed) Saharan dust. Wiley, Chichester, pp. 187–193

Yaalon DH, Ginzbourg D (1966) Sedimentary characteristics and climatic analysis of easterly dust storms in the Negev (Israel). Sedimentology 6:315–322

Yabuki S, et al. (2005) The characteristics of atmospheric aerosol at Aksu, and Asian dust-source region of North-West China: a summary of observations over the three years from March 2001 to April 2004. J Meteorol Soc Jpn 83A:45–72

Yadav S, Rajamani V (2004) Geochemistry of aerosols of northwestern part of India adjoining the Thar Desert. Geochim Cosmochim Acta 68:1975–1988

Yamada K (2004) Last 40 ka climate changes as deduced from the lacustrine sediments of Lake Biwa, central Japan. Quat Int 123/125:43–50

Yan H, Wang S, Wang C, Zhang G, Patel N (2005) Losses of soil organic carbon under wind erosion in China. Global Change Biol 11:828–840

Yang C-Y, Chen Y-S, Chiu H-F, Goggins WB (2005) Effects of Asian dust storm events on daily stroke admissions in Taipei, Taiwan. Environ Res 99:79–84

Yang G, Xiao H, Tuo W (2001) Black windstorms in northwest China: a case study of thre strong sand-dust storm of May 5th 1993. In: Yang Y, Squires V, Qi L (eds) Global alarm: dust and sandstorms in the world's drylands. UNESCO, Bangkok, pp. 49–73

Yang SL, Ding ZL (2004) Comparison of particle size characteristics of the tertiary 'red clay' and Pleistocene loess in the Chinese Loess Plateau: implications for origin and sources of the 'red clay'. Sedimentology 51:77–93

Yatagai S, Takemura K, Naruse T, Kitagawa H, Fukusawa H, Kim M-H, Yasuda Y (2002) Monsoon changes and eolian dust deposition over the past 30,000 years in Cheju Island, Korea. Trans Jpn Geomorphol Union 23/25:821–831

Yoshino M (1992) Wind and rain in the desert region of Xinjiang, northwest China. Erdkunde 46:203–216

Youngsin C, Lim J-Y (2003) The recent characteristics of Asian dust and haze events in Seoul, Korea. Meteorol Atmos Phys Spec Iss Air Qual 10, doi:10.1007/s00703-003-0067-2

Yu H, Dickinson RE, Chin M, Kaufman YJ, Holben BN, Geogdzhayev IV, Mischensko MI (2003) Annual cycle of global distributions of aerosol optical depth from integration of MODIS retrievals and GOCART model simulations. J Geophys Res 108D, doi:10.1029/2002JD002717

Yuan CS, Sau CC, Chen MC, Hunag MH, Chang SW, Lin YC, Lee CG (2004) Mass concentration and size-resolved chemical composition of atmospheric aerosols sampled at the Pescadores Islands during Asian dust storm periods in the years of 2001 and 2002. Terr Atmos Ocean Sci 15:857–879

Yunnie P (2002) Fighting with Popski's private army. Greenhill Books, London

Zaady E, Offer ZY, Shachak M (2001) The content and contribution of deposited aeolian organic matter in a dry land ecosystem of the Negev desert, Israel. Atmos Environ 35:769–776

Zaizen Y, Ikegami M, Okada K, Makino Y (1995) Aerosol concentration observed at Zhangye in China. J Meteorol Soc Jpn 73:891–897

Zakharov PS (1966) The characteristics and geographical distribution of dust storms (in Russian). Meteor Klimatol Gidrol 1966:19–22

Zarate MA (2003) Loess of southern South America. Quat Sci Rev 22:1987–2006

Zender CS, Kwon EY (2005) Regional contrasts in dust emission responses to climate. J Geophys Res 110:D13201, doi:10.1029/2004JD005501

Zender CS, Newman D, Torres O (2003) Spatial heterogeneity in aeolian erodibility: uniform, topographic, geomorphic and hydrologic hypotheses. J Geophys Res 108D:4543

Zhang DE (1985) Meteorological characteristics of dust fall in China since the historic times. In: Liu TS (ed) Quaternary geology and environment of China. China Ocean Press, Beijing, pp. 101–106

Zhang J, Christopher SA (2003) Longwave radiative forcing of Saharan dust aerosols estimated from MODIS, MISR and CERES observations on terra. Geophys Res Lett 30:ASC3-1-4

Zhang R, Wang M, Zhang X, Zhu G (2003) Analysis of the chemical and physical properties of particles in a dust storm in spring in Beijing. Powder Technol 137:77–82

Zhang XY, Arimoto R, An ZS (1997) Dust emission from Chinese desert sources linked to variations in atmospheric circulation. J Geophys Res 102(D23): 28041–28047

Zhang XY, Arimoto R, Zhu GH, Chen T, Zhang GY (1998) Concentration, size-distribution and deposition of mineral aerosol over Chinese desert regions. Tellus 50B:317–330

Zhang XY, Gong SL, Zhao TL, Arimoto R, Wang YQ, Zhou ZJ (2003) Sources of Asian dust and role of climate change versus desertification in Asian dust emission. Geophys Res Lett 30:2272

Zhang Z, Zhao M, Eglinton G, Lu H, Huang C-Y (2006) Leaf wax lipids as paleovegetational and paleoenvironmental proxies for the Chinese Loess Plateau over the last 170 kyr. Quat Sci Rev (in press)

Zhao TL, Gong SL, Zhang XY, McKendry IG (2003) Modeled size-aggregated wet and dry deposition budgets of soil dust aerosol during ACE-Asia 2001: implications for trans-Pacific transport. J Geophys Res 108:D23, doi:10.1029/2002JD003363

Zhong SL, Zhu YH, Willems H, Xu JL (2003) On the use of coccoliths to reconstruct dust soyurce areas and transport routes: an example from the Malan Loess of the Penglai district, Northwestern Shandong Peninsula, China. Courr Forschungsinst Senckenberg 244:129–135

Zhou LP, Dodonov AE, Shackleton NJ (1995) Thermoluminescence dating of the Orkutsay loess section in Tashkent Region, Uzbekistan, Central Asia. Quat Sci Rev 14:721–730

Zhou XKK, Zhai PMM (2004) Relationship between vegetation coverage and spring dust storms over northern China. J Geophys Res 109D:D03104

Zhou Z, Zhang G (2003) Typical severe dust storms in northern China during 1954–2002. Chin Sci Bull 48, doi: 10.1360/03wd0029

Zhu XR, Prospero JM, Miller FJ (1997) Daily variability of soluble Fe (II) and soluble total Fe in North African dust in the trade winds at Barbados. J Geophys Res 102(D17):21297–21305

Zhu Z (1984) Aeolian landforms in the Taklimakan Desert. In: El-Baz F (ed) Deserts and arid lands. Nijhoff, The Hague, pp. 133–144

Zinck JA, Sayago JM (2001) Climatic periodicity during the late Pleistocene from a loess-paleosol sequence in northwest Argentina. Quat Int 78:11–16

Zurek, Martin LJ (1993) Interannual variability of planet-encircling dust storms on Mars. J Geophys Res 98:3247–3259

Index